LONDON MATHEMATICAL SOCIETY LECTURE NOTE SERIES

Managing Editor: Professor J.W.S. Cassels, Department of Pure Mathematics and Mathematical Statistics, University of Cambridge, 16 Mill Lane, Cambridge CB2 1SB, England

The titles below are available from booksellers, or, in case of difficulty, from Cambridge University Press.

34	Representation theory of Lie groups, M.F. ATIYAH *et al*
46	p-adic analysis: a short course on recent work, N. KOBLITZ
50	Commutator calculus and groups of homotopy classes, H.J. BAUES
59	Applicable differential geometry, M. CRAMPIN & F.A.E. PIRANI
66	Several complex variables and complex manifolds II, M.J. FIELD
69	Representation theory, I.M. GELFAND *et al*
76	Spectral theory of linear differential operators and comparison algebras, H.O. CORDES
77	Isolated singular points on complete intersections, E.J.N. LOOIJENGA
83	Homogeneous structures on Riemannian manifolds, F. TRICERRI & L. VANHECKE
86	Topological topics, I.M. JAMES (ed)
87	Surveys in set theory, A.R.D. MATHIAS (ed)
88	FPF ring theory, C. FAITH & S. PAGE
89	An F-space sampler, N.J. KALTON, N.T. PECK & J.W. ROBERTS
90	Polytopes and symmetry, S.A. ROBERTSON
92	Representation of rings over skew fields, A.H. SCHOFIELD
93	Aspects of topology, I.M. JAMES & E.H. KRONHEIMER (eds)
94	Representations of general linear groups, G.D. JAMES
95	Low-dimensional topology 1982, R.A. FENN (ed)
96	Diophantine equations over function fields, R.C. MASON
97	Varieties of constructive mathematics, D.S. BRIDGES & F. RICHMAN
98	Localization in Noetherian rings, A.V. JATEGAONKAR
99	Methods of differential geometry in algebraic topology, M. KAROUBI & C. LERUSTE
100	Stopping time techniques for analysts and probabilists, L. EGGHE
104	Elliptic structures on 3-manifolds, C.B. THOMAS
105	A local spectral theory for closed operators, I. ERDELYI & WANG SHENGWANG
107	Compactification of Siegel moduli schemes, C-L. CHAI
108	Some topics in graph theory, H.P. YAP
109	Diophantine analysis, J. LOXTON & A. VAN DER POORTEN (eds)
110	An introduction to surreal numbers, H. GONSHOR
113	Lectures on the asymptotic theory of ideals, D. REES
114	Lectures on Bochner-Riesz means, K.M. DAVIS & Y-C. CHANG
115	An introduction to independence for analysts, H.G. DALES & W.H. WOODIN
116	Representations of algebras, P.J. WEBB (ed)
118	Skew linear groups, M. SHIRVANI & B. WEHRFRITZ
119	Triangulated categories in the representation theory of finite-dimensional algebras, D. HAPPEL
121	Proceedings of *Groups - St Andrews 1985*, E. ROBERTSON & C. CAMPBELL (eds)
122	Non-classical continuum mechanics, R.J. KNOPS & A.A. LACEY (eds)
125	Commutator theory for congruence modular varieties, R. FREESE & R. MCKENZIE
126	Van der Corput's method of exponential sums, S.W. GRAHAM & G. KOLESNIK
128	Descriptive set theory and the structure of sets of uniqueness, A.S. KECHRIS & A. LOUVEAU
129	The subgroup structure of the finite classical groups, P.B. KLEIDMAN & M.W. LIEBECK
130	Model theory and modules, M. PREST
131	Algebraic, extremal & metric combinatorics, M-M. DEZA, P. FRANKL & I.G. ROSENBERG (eds)
132	Whitehead groups of finite groups, ROBERT OLIVER
133	Linear algebraic monoids, MOHAN S. PUTCHA
134	Number theory and dynamical systems, M. DODSON & J. VICKERS (eds)
135	Operator algebras and applications, 1, D. EVANS & M. TAKESAKI (eds)
136	Operator algebras and applications, 2, D. EVANS & M. TAKESAKI (eds)
137	Analysis at Urbana, I, E. BERKSON, T. PECK, & J. UHL (eds)
138	Analysis at Urbana, II, E. BERKSON, T. PECK, & J. UHL (eds)
139	Advances in homotopy theory, S. SALAMON, B. STEER & W. SUTHERLAND (eds)
140	Geometric aspects of Banach spaces, E.M. PEINADOR & A. RODES (eds)
141	Surveys in combinatorics 1989, J. SIEMONS (ed)
144	Introduction to uniform spaces, I.M. JAMES
145	Homological questions in local algebra, JAN R. STROOKER
146	Cohen-Macaulay modules over Cohen-Macaulay rings, Y. YOSHINO
147	Continuous and discrete modules, S.H. MOHAMED & B.J. MÜLLER
148	Helices and vector bundles, A.N. RUDAKOV *et al*
149	Solitons, nonlinear evolution equations and inverse scattering, M. ABLOWITZ & P. CLARKSON
150	Geometry of low-dimensional manifolds 1, S. DONALDSON & C.B. THOMAS (eds)
151	Geometry of low-dimensional manifolds 2, S. DONALDSON & C.B. THOMAS (eds)
152	Oligomorphic permutation groups, P. CAMERON
153	L-functions and arithmetic, J. COATES & M.J. TAYLOR (eds)
154	Number theory and cryptography, J. LOXTON (ed
155	Classification theories of polarized varieties, TAKAO FUJITA
156	Twistors in mathematics and physics, T.N. BAILEY & R.J. BASTON (eds)

158 Geometry of Banach spaces, P.F.X. MÜLLER & W. SCHACHERMAYER (eds)
159 Groups St Andrews 1989 volume 1, C.M. CAMPBELL & E.F. ROBERTSON (eds)
160 Groups St Andrews 1989 volume 2, C.M. CAMPBELL & E.F. ROBERTSON (eds)
161 Lectures on block theory, BURKHARD KÜLSHAMMER
162 Harmonic analysis and representation theory, A. FIGA-TALAMANCA & C. NEBBIA
163 Topics in varieties of group representations, S.M. VOVSI
164 Quasi-symmetric designs, M.S. SHRIKANDE & S.S. SANE
165 Groups, combinatorics & geometry, M.W. LIEBECK & J. SAXL (eds)
166 Surveys in combinatorics, 1991, A.D. KEEDWELL (ed)
167 Stochastic analysis, M.T. BARLOW & N.H. BINGHAM (eds)
168 Representations of algebras, H. TACHIKAWA & S. BRENNER (eds)
169 Boolean function complexity, M.S. PATERSON (ed)
170 Manifolds with singularities and the Adams-Novikov spectral sequence, B. BOTVINNIK
171 Squares, A.R. RAJWADE
172 Algebraic varieties, GEORGE R. KEMPF
173 Discrete groups and geometry, W.J. HARVEY & C. MACLACHLAN (eds)
174 Lectures on mechanics, J.E. MARSDEN
175 Adams memorial symposium on algebraic topology 1, N. RAY & G. WALKER (eds)
176 Adams memorial symposium on algebraic topology 2, N. RAY & G. WALKER (eds)
177 Applications of categories in computer science, M. FOURMAN, P. JOHNSTONE, & A. PITTS (eds)
178 Lower K- and L-theory, A. RANICKI
179 Complex projective geometry, G. ELLINGSRUD et al
180 Lectures on ergodic theory and Pesin theory on compact manifolds, M. POLLICOTT
181 Geometric group theory I, G.A. NIBLO & M.A. ROLLER (eds)
182 Geometric group theory II, G.A. NIBLO & M.A. ROLLER (eds)
183 Shintani zeta functions, A. YUKIE
184 Arithmetical functions, W. SCHWARZ & J. SPILKER
185 Representations of solvable groups, O. MANZ & T.R. WOLF
186 Complexity: knots, colourings and counting, D.J.A. WELSH
187 Surveys in combinatorics, 1993, K. WALKER (ed)
188 Local analysis for the odd order theorem, H. BENDER & G. GLAUBERMAN
189 Locally presentable and accessible categories, J. ADAMEK & J. ROSICKY
190 Polynomial invariants of finite groups, D.J. BENSON
191 Finite geometry and combinatorics, F. DE CLERCK et al
192 Symplectic geometry, D. SALAMON (ed)
193 Computer algebra and differential equations, E. TOURNIER (ed)
194 Independent random variables and rearrangement invariant spaces, M. BRAVERMAN
195 Arithmetic of blowup algebras, WOLMER VASCONCELOS
196 Microlocal analysis for differential operators, A. GRIGIS & J. SJÖSTRAND
197 Two-dimensional homotopy and combinatorial group theory, C. HOG-ANGELONI,
    W. METZLER & A.J. SIERADSKI (eds)
198 The algebraic characterization of geometric 4-manifolds, J.A. HILLMAN
199 Invariant potential theory in the unit ball of C^n, MANFRED STOLL
200 The Grothendieck theory of dessins d'enfant, L. SCHNEPS (ed)
201 Singularities, JEAN-PAUL BRASSELET (ed)
202 The technique of pseudodifferential operators, H.O. CORDES
203 Hochschild cohomology of von Neumann algebras, A. SINCLAIR & R. SMITH
204 Combinatorial and geometric group theory, A.J. DUNCAN, N.D. GILBERT & J. HOWIE (eds)
205 Ergodic theory and its connections with harmonic analysis, K. PETERSEN & I. SALAMA (eds)
206 An introduction to noncommutative differential geometry and its physical applications, J. MADORE
207 Groups of Lie type and their geometries, W.M. KANTOR & L. DI MARTINO (eds)
208 Vector bundles in algebraic geometry, N.J. HITCHIN, P. NEWSTEAD & W.M. OXBURY (eds)
209 Arithmetic of diagonal hypersurfaces over finite fields, F.Q. GOUVÊA & N. YUI
210 Hilbert C*-modules, E.C. LANCE
211 Groups 93 Galway / St Andrews I, C.M. CAMPBELL et al
212 Groups 93 Galway / St Andrews II, C.M. CAMPBELL et al
214 Generalised Euler-Jacobi inversion formula and asymptotics beyond all orders, V. KOWALENKO,
    N.E. FRANKEL, M.L. GLASSER & T. TAUCHER
215 Number theory, S. DAVID (ed)
216 Stochastic partial differential equations, A. ETHERIDGE (ed)
217 Quadratic forms with applications to algebraic geometry and topology, A. PFISTER
218 Surveys in combinatorics, 1995, PETER ROWLINSON (ed)
220 Algebraic set theory, A. JOYAL & I. MOERDIJK
221 Harmonic approximation, S.J. GARDINER
222 Advances in linear logic, J.-Y. GIRARD, Y. LAFONT & L. REGNIER (eds)
223 Analytic semigroups and semilinear initial boundary value problems, KAZUAKI TAIRA
224 Computability, enumerability, unsolvability, S.B. COOPER, T.A. SLAMAN & S.S. WAINER (eds)
226 Novikov conjectures, index theorems and rigidity I, S. FERRY, A. RANICKI & J. ROSENBERG (eds)
227 Novikov conjectures, index theorems and rigidity II, S. FERRY, A. RANICKI & J. ROSENBERG (eds)
228 Ergodic theory of Z^d actions, M. POLLICOTT & K. SCHMIDT (eds)
229 Ergodicity for infinite dimensional systems, G. DA PRATO & J. ZABCZYK
230 Prolegomena to a middlebrow arithmetic of curves of genus 2, J.W.S. CASSELS & E.V. FLYNN
231 Semigroup theory and its applications, K.H. HOFMANN & M.W. MISLOVE (eds)

London Mathematical Society Lecture Note Series. 230

Prolegomena to a Middlebrow Arithmetic of Curves of Genus 2

J.W.S. Cassels
University of Cambridge

E.V. Flynn
University of Liverpool

CAMBRIDGE
UNIVERSITY PRESS

CAMBRIDGE UNIVERSITY PRESS
Cambridge, New York, Melbourne, Madrid, Cape Town, Singapore, São Paulo

Cambridge University Press
The Edinburgh Building, Cambridge CB2 2RU, UK

Published in the United States of America by Cambridge University Press, New York

www.cambridge.org
Information on this title: www.cambridge.org/9780521483704

First published 1996

A catalogue record for this publication is available from the British Library

ISBN-13 978-0-521-48370-4 paperback
ISBN-10 0-521-48370-0 paperback

Transferred to digital printing 2005

Contents

Chapters are divided into sections. Some topics occupy several sections. We give a name to a section and list it here only when it starts a new topic.

Foreword ix

Background and conventions
   0. Introduction xi
   1. Algebraically closed ground field xi
   2. Perfect ground field xii
   Appendix. Finite-dimensional Galois modules xiii

Chapter 1. Curves of genus 2
   1. Canonical form 1
   2. Group law 3
   3. The form is canonical 4
   4. Plane quartics 4

Chapter 2. Construction of the jacobian
   0. Introduction 6
   1. Construction of a basis 6
   3. Local behaviour 9
   4. Formal group law 10
   Appendix I. Blowing up and blowing down 12
   Appendix II. Change of coordinates on $\mathcal{C}$ 13

Chapter 3. The Kummer surface
   0. Introduction 17
   1. Construction of the surface. Nodes 18
   2. Addition of 2-division points 20
   3. Commutativity properties 22
   4. The biquadratic forms 23
   6. A numerical example 24
   7. The tropes 25
   8. The jacobian as double cover 26
   9. The group law on the jacobian 27
   10. Recovery of the curve 29

Chapter 4. The dual of the Kummer
0. Introduction 31
1. Description of Pic^3. 31
3. $\mathcal{K}^*$ is the projective dual of $\mathcal{K}$ 34
4. Tangency 36
5. Explicit dualities 37
Appendix. Rational divisor classes with no rational divisor 39

Chapter 5. Weddle's surface
0. Introduction 40
1. Symmetroid and Jacobian 41
2. A special case 42
4. Duality 44

Chapter 6. $\mathfrak{G}/2\mathfrak{G}$
0. Introduction 47
1. The homomorphism 48
2. The kernel 50
5. When is $\mathfrak{W} \subset 2\mathfrak{G}$? 55
6. The Kummer viewpoint 57
8. The norm of ρ 59
9. A pathology 61
10. The quintic case 61

Chapter 7. The jacobian over local fields. Formal groups
0. Introduction 63
1. Computing the formal group 63
2. General properties of formal groups 67
3. The reduction map 69
4. Torsion in the kernel of reduction 70
5. The order of $\mathfrak{G}/2\mathfrak{G}$ when k is a finite extension of $\mathbb{Q}_p$, $p \neq \infty$. 70
6. The order of $\mathfrak{G}/2\mathfrak{G}$ when $k = \mathbb{R}$ or $\mathbb{C}$ 73

Chapter 8. Torsion
0. Introduction 75
1. Computing the group law 75
2. Computing the torsion of a given jacobian 78
3. Searching for large rational torsion 82

Chapter 9. The isogeny. Theory

0. Introduction 88
4. The Kummer formulation 91
5. Statement of results 92
6. Motivation 93
7. The isogeny 95
8. Generation by radicals 96
9. Consequences for Mordell-Weil 97
10. Isogeny for the curve 98

Chapter 10. The isogeny. Applications

0. Introduction 101
1. The isogeny on the jacobian variety 101
2. An injection on $\widehat{\mathfrak{G}}/\phi(\mathfrak{G})$ 104
3. Homogeneous spaces $J^{\phi}_{d_1,d_2}$ and norm spaces $L^{\phi}_{d_1,d_2}$ 109
4. Computing $\#\widehat{\mathfrak{G}}/\phi(\mathfrak{G}) \cdot \#\mathfrak{G}/\hat{\phi}(\widehat{\mathfrak{G}})$ when $k = \mathbb{Q}_p$ 110

Chapter 11. Computing the Mordell-Weil group

0. Introduction 113
1. The Weak Mordell-Weil Theorem 113
2. Performing 2-descent without using homogeneous spaces 116
3. A worked example of complete 2-descent 121
4. A worked example of descent via isogeny 125
5. Large rank 130

Chapter 12. Heights

0. Introduction 133
1. A height function on $\mathfrak{G}$ 133
2. The Mordell-Weil Theorem 137
3. A computational improvement 140

Chapter 13. Rational points. Chabauty's Theorem

0. Introduction 143
1. Chabauty's Theorem 143
2. A worked example 148

Chapter 14. Reducible jacobians

0. Introduction 154
1. The straightforward case 154
2. An awful warning 156
3. The reverse process 157
4. Trying to show that a given jacobian is simple 157

Chapter 15. The endomorphism ring

0. Introduction 160
1. A few examples 160
2. Complex and real multiplication 163

Chapter 16. The desingularized Kummer

0. Introduction 165
1. The surface 165
2. Elementary properties 167
3. Equivalence with $\mathcal{K}$ 168
4. Equivalence with $\mathcal{K}^*$ 170
5. Comparison of maps 172
6. Further possible developments 172

Chapter 17. A neoclassical approach

0. Introduction 174
1. Proof of theorem 176
2. The Kummer surface 179
3. A special case 181
4. The construction 183

Chapter 18. Zukunftsmusik

0. Introduction 186
1. Framework 186
2. Principal homogeneous spaces 186
3. Duality 186
4. Canonical height 186
5. Second descents 187
6. Other genera 187
7. A challenge of Serre's 187

Appendix I. MAPLE programs 190

Appendix II. Files available by anonymous ftp 204

Bibliography 207

Index rerum et personarum 219

Foreword

The arithmetic (= number theory) of curves of genus 0 is well understood. For genus 1, there is a rich body of theory and conjecture, and in recent years notable success has been achieved in transforming the latter into the former. For curves of higher genus there is a rich body of theory with some spectacular successes (e.g. Faltings' proof of 'Mordell's Conjecture'), but our command is still rudimentary. For genus 1, a concrete question about an individual curve can usually be answered by one means or another: for higher genus, this is far from the case.

For curves of genus 0 and 1 the road between general theory and particular cases is no one-way street. Numerous individual cases were investigated by amateurs as well as as by distinguished mathematicians such as Diophantos, Fermat, Euler, Sylvester, Mordell, Selmer, Birch and Swinnerton-Dyer. Regularities which emerged, sometimes quite unexpectedly, suggested theorems, which could sometimes be proved. The new theorems suggested new questions. For higher genus, existing theory is notoriously unadapted to the study of individual curves, and few have been elucidated. What is needed is a corpus of explicit concrete cases and a middlebrow arithmetic theory which would provide both a practicable means to obtain them and a framework to understand any unexpected regularities.

Such a theory will, of course, draw freely from the existing body of knowledge, but the emphasis on feasibility gives new perspectives. Classical geometers worked over an algebraically closed field, usually the complexes. Modern geometers take account of the ground field, but regard a field extension as a cheap manoeuvre. For practical computations it is desperately expensive: the theory we seek must avoid them at almost all costs.

A natural place to begin is with a curve of genus 2 defined over the rationals and a natural first problem is to determine its Mordell-Weil group (or at least the rank). General theory suggests an approach through the (arithmetic) jacobian, the arithmetic analogue of the complex manifold studied in the 19th century. Although it figures prominently in the general theory, we appear to have been the first to describe it explicitly, as a surface in 15-dimensional projective space. It is an inconveniently large object, but fortunately much information is retained in its Kummer surface. These are surfaces of degree 4, which were also the subject of intensive 19th century study over the reals and complexes.

Our first attack on the Mordell-Weil group was a simple-minded generalization of an algorithm for elliptic curves. The study of the classical geometrical literature gave a deeper understanding of the algorithm and also new perspectives on the geometric theory. The algorithm is still not easy. In appropriate cases, however, there is an isogeny of degree 4 analogous to that of degree 2 for elliptic curves which, where it exists, permits the determination of the rank on a production-line basis.¶

A less sophisticated, but deeper, problem is to ask for all the rational points on a curve of genus 2. The number is finite by Faltings' Theorem, but the proof gives no hope of a reasonable algorithm. An old theorem of Chabauty states (as a special case) that the number is finite if the Mordell-Weil rank is (0 or) 1. As a theorem, it is superseded by Faltings, but the proof gives a feasible approach. Until recently, no actual example was known. The first was given by† Gordon & Grant (1993), and now there are many.

Some chapters do not require everything that goes before. In particular, the computational Chapters 8, 10, 11, 12, 13 do not depend on Chapters 4, 5.

Even with the emphasis on feasibility, the manipulations would have daunted a Gauss or a Salmon. Our investigations would have been impossible without the resources of computer algebra. We have not attempted to reproduce the larger formulae in the text, but have made them available‡ in machine-readable form by ftp. We give in Appendix I programs which enable the reader to check our assertions and to produce formulae for herself. They are in the computer language MAPLE, the algebra package we have mainly used, and are heavily annotated: a reader who prefers another package should have no difficulty in adapting them. The MAPLE programs are also available by ftp.

This volume is a progress report on a project which has lasted, on and off, more than ten years. We are grateful to SERC and its successor EPSRC for grants to JWSC which supported EVF for four years. The text has benefited from Ed Schaefer's criticism of an earlier draft. The canonical disclaimer applies.

¶ As for elliptic curves, the algorithm is not effective in the logical sense. But it usually works.

† References are cited by name(s) and date, with a possible further letter.

‡ See Appendix II.

Background and conventions

0 Introduction. In writing we have had in mind a beginning graduate student with some exposure to algebraic geometry.¶ Here we indicate our point of view. We fix some notation and recall some definitions and background. These are illustrated by a simple computation.

We shall be working over a ground field such as the rationals $\mathbb{Q}$ or the p-adics $\mathbb{Q}_p$. First, however, we look at an algebraically closed field.

1 Algebraically closed ground field. Let $\bar{k}$ be an algebraically closed field, for simplicity of characteristic 0, and let $\mathcal{B}$ be a complete curve defined over $\bar{k}$. For example $\mathcal{B}$ may be a nonsingular curve in a projective space defined by equations with coefficients in $\bar{k}$. A *divisor* $\mathfrak{A} = \sum_{\mathfrak{a}} n_{\mathfrak{a}} \mathfrak{a}$ on $\mathcal{B}$ is an element of the free abelian group Div on (symbols for) the points on $\mathcal{B}$ defined over $\bar{k}$. A point $\mathfrak{a}$ of $\mathcal{B}$ has a multiplicity $n_{\mathfrak{a}}$ in $\mathfrak{A}$ which is an integer, and is 0 for all except finitely many $\mathfrak{a}$. The set of points with nonzero multiplicity in a divisor is its *support*. The *degree* of a divisor is the sum $\sum_{\mathfrak{a}} n_{\mathfrak{a}}$ of the multiplicities of the points.

A nonzero function f on $\mathcal{B}$ determines a divisor $[f]$ as follows. The multiplicity† of a point $\mathfrak{a}$ in $[f]$ is the order to which f vanishes at $\mathfrak{a}$ measured in terms of a *local uniformizer*: so the multiplicity is negative if $\mathfrak{a}$ is a pole. A divisor of the form $[f]$ is a *principal divisor*. Principal divisors have degree 0.

The principal divisors are a subgroup of Div. Two divisors differing by a principal divisor are *linearly equivalent* or in the same *divisor class*. The divisor classes form the *Picard group* Pic. Pic inherits a degree from Div. The set of elements of Pic of degree j is denoted by Pic^j. Clearly Pic^0 is a subgroup of Pic.

¶ e.g. Chapters I, II of Silverman (1986). We shall also assume that the reader is familiar with the idea of a field with a valuation, in particular the p-adic numbers $\mathbb{Q}_p$ [see Chapter 2 of Cassels (1991) or the first few chapters of Cassels (1986)]. It would be helpful for the reader to have had a brief introduction to elliptic curves; either Cassels (1991) or Chapters III, IV, VII, VIII of Silverman (1986) would be more than sufficient.

† A Mickey Mouse example follows.

Similarly we can define the divisor of a *differential* hdf, where f, h are functions, in terms of local uniformizers. The divisors of differentials are all in the same divisor class, the *canonical class*. The degree of the canonical class is $2g - 2$, where g is the *genus*. A divisor $\mathfrak{A}$ is *effective* if all the points of its support have nonnegative multiplicity. There is a partial order $\succ$ on Div: by definition $\mathfrak{A} \succ \mathfrak{B}$ if $\mathfrak{A} - \mathfrak{B}$ is effective. For given divisor $\mathfrak{A}$, the set of functions f with¶ $[f] \succ -\mathfrak{A}$ together with $f = 0$ form a vector space over $\bar{k}$. It has a finite dimension given by the *Riemann-Roch Theorem*, which we do not rehearse.

Still working over an algebraically closed field $\bar{k}$, we suppose that the characteristic is not 2 and take as an example the curve $\mathcal{C}$ given in the affine plane by

$$Y^2 = \prod_{j=1}^{6}(X - \theta_j).$$

It is not complete: there are two points $\mathfrak{b}_1$, $\mathfrak{b}_2$ 'at infinity'. The points $\mathfrak{a}_j = (\theta_j, 0)$ are the *Weierstrass points*. Then Y is a local uniformizer at the $\mathfrak{a}_j$ and X^{-1} is a local uniformizer at the $\mathfrak{b}_j$. Otherwise, $(X - x)$ is a local uniformizer at (x, y). Hence the divisor of dX is†

$$-2\mathfrak{b}_1 - 2\mathfrak{b}_2 + \sum \mathfrak{a}_j,$$

which is by definition in the canonical class. It has degree

$$6 - 2 - 2 = 2 = 2g - 2,$$

so the genus g is 2. Since

$$[Y] = -3\mathfrak{b}_1 - 3\mathfrak{b}_2 + \sum \mathfrak{a}_j,$$

it follows that $\mathfrak{b}_1 + \mathfrak{b}_2$, the divisor of poles of X, is in the canonical class.

2   Perfect ground field. One can develop algebraic geometry from scratch over a general ground field k [e.g. Chevalley (1951)]. We are concerned only with k of characteristic 0 or, very occasionally, of finite cardinality, so a less sophisticated approach suits us better. Let $\bar{k}$ be the algebraic

---

¶ The negative sign here is traditional.

† The multiplicity $n_\mathfrak{p}$ of $\mathfrak{p}$ is the order of d$X/dt_\mathfrak{p}$ at $\mathfrak{p}$, where $t_\mathfrak{p}$ is a local uniformizer: that is $t_\mathfrak{p}^{-n_\mathfrak{p}} dX/dt_\mathfrak{p}$ has neither a zero nor a pole at $\mathfrak{p}$.

closure of k. Then $\mathrm{Gal} = \mathrm{Gal}\,(\bar{k}/k)$ acts on all the objects in the algebraic geometry over $\bar{k}$. We say that a divisor, etc. is *defined over* k if it is fixed under the action of Galois. If there is only one ground field k of interest at the time, we may say *rational* instead of 'defined over k'. (But we have to avoid the phrase 'a rational curve' since this is established as a synonym of 'a curve of genus 0'.) The only slightly tricky point to note is that if a divisor is defined over k and principal, then it is the divisor of a function defined over k. This is a special case of a general principle proved in the following Appendix. On the other hand, and this will turn out to be crucial, a rational divisor class does not necessarily contain a rational divisor, see the Appendix to Chapter 4.

Appendix

Finite-dimensional Galois modules

The following theorem is often useful in deducing results over k from those over $\bar{k}$ [Cartier (1960)].

THEOREM. *Let $\mathfrak{M}$ be a finite-dimensional $\bar{k}$-module. Suppose given an action of Gal on $\mathfrak{M}$ compatible with the $\bar{k}$-structure. We shall say that $\mu \in \mathfrak{M}$ is defined over an extension K of k if μ is fixed by every $\sigma \in \mathrm{Gal}$ which leaves K elementwise fixed. Suppose that every element of $\mathfrak{M}$ is defined over some finite extension of k. Then $\mathfrak{M}$ has a $\bar{k}$-basis consisting of elements defined over k.*

Note. The 1-dimensional case is the standard generalization of Hilbert 90 under a light disguise, cf. Serre (1962), p. 159.

Proof. We denote by $\mathfrak{M}_k$ the set of elements of $\mathfrak{M}$ defined over k. It inherits a structure as k-vector space from $\mathfrak{M}$.

(i) $\mathfrak{M}$ *is $\bar{k}$-spanned by the elements of $\mathfrak{M}_k$.* Let $\mu \in \mathfrak{M}$, so μ is defined over a finite extension K of k, which without loss of generality may be supposed to be normal. Let $\omega_1, \ldots, \omega_n$ be a basis of K/k and let $\tau_1, \ldots, \tau_n$ be representatives in Gal of the Galois group of K/k. Put

$$
\begin{aligned}
m_j &= \sum_{i=1}^{n} \tau_i(\omega_j \mu) \\
&= \sum_{i=1}^{n} (\tau_i \omega_j)(\tau_i \mu) \qquad (1 \leqslant j \leqslant n).
\end{aligned}
$$

Clearly $m_j \in \mathfrak{M}_k$. Since $\det_{i,j}(\tau_i \omega_j) \neq 0$ by a basic theorem of Galois theory, μ is a $\bar{k}$-linear combination of the m_j.

(ii) *If m_j $(1 \leqslant j \leqslant J)$ in $\mathfrak{M}_k$ are linearly dependent over $\bar{k}$, then they are already linearly dependent over k.* The proof is similar. Suppose that $\sum \lambda_j m_j = 0$, where the λ_j are not all 0 and, without loss of generality, are all in some finite normal extension K of k. Then by multiplying the relation by the elements of a basis of K/k and taking traces, we get linear relations between the m_j with coefficients in k, not all 0.

(iii) We can now complete the proof. By hypothesis $\mathfrak{M}$ is $\bar{k}$-spanned by a finite set of elements, and so, by (i), it is $\bar{k}$-spanned by a finite set $\mathcal{S} \subset \mathfrak{M}_k$. Let $\mathcal{S}_0$ be a maximal k-independent subset of $\mathcal{S}$. Then $\mathcal{S}_0$ is $\bar{k}$-independent by (ii), and so is the required $\bar{k}$-basis of $\mathfrak{M}$.

Chapter 1

Curves of genus 2

1   Canonical form. We shall normally suppose that the characteristic¶ of the ground field is not 2 and consider curves $\mathcal{C}$ of genus 2 in the shape

$$\mathcal{C}: \quad Y^2 = F(X), \tag{1.1.1}$$

where

$$F(X) = f_0 + f_1 X + \ldots + f_6 X^6 \in k[X] \tag{1.1.2}$$

is of degree 6 and has no multiple factors. As we shall see in a moment, every curve of genus 2 defined over k is birationally equivalent over k to a curve of this type, which is unique up to a fractional linear transformation of X, and associated transformation of Y,

$$X \longmapsto (aX + b)/(cX + d), \quad Y \longmapsto eY/(cX + d)^3, \tag{1.1.3}$$

where

$$a, b, c, d \in k, \quad ad - bc \neq 0, \quad e \in k^*. \tag{1.1.4}$$

When k is algebraically closed, we can always use a transformation of this kind to ensure that F is of degree 5:† this is what is almost always done in the classical theory. From our point of view, this is very special. We shall assume, by using a transformation (3)‡ if necessary, that F is of precise

---

¶ A more systematic treatment would require consideration of finite fields of characteristic 2, since these arise e.g. by reduction mod 2 when $k = \mathbb{Q}$. We shall not need it. The modifications required for characteristic 2 are left as optional exercises for the reader.

† This is possible precisely when the sextic $F(X)$ has a root $\alpha \in k$. Then we can use the map $X \mapsto 1/(X-\alpha)$, $Y \mapsto Y/(X-\alpha)^3$ from $Y^2 = F(X)$ to a curve $Y^2 = $ (quintic in X). For example, $X \mapsto 1/(X-2)$, $Y \mapsto Y/(X-2)^3$ maps $Y^2 = X^6 - 64$ to $Y^2 = (2X+1)^6 - 64X^6 = $ (quintic in X) [The second equation is obtained by replacing X by $(2X+1)/X$, Y by Y/X^3 in the first equation, and then multiplying through by X^6].

‡ Displayed formulae are referred to by a number triple. For example, (2.3.4) is display 4 in Section 3 of Chapter 2. It is cited as (4) within Section 3 of Chapter 2, as (3.4) in the rest of Chapter 2, and as (2.3.4) outside Chapter 2.

degree 6.¶

If we attempt to obtain a complete curve from the affine curve (1) by replacing X, Y by $X/Z, Y/Z^3$, we get an unpleasant singularity at $Z = 0$. We will get round this in one of two ways. The first is *à la Weil* to use a transformation (3) to ensure that none of the points we are interested in is 'at infinity'. Alternatively, we can use the complete nonsingular model

$$Y^2 = f_0 X_0^2 + f_1 X_0 X_1 + f_2 X_1^2 + f_3 X_1 X_2 + f_4 X_2^2 + f_5 X_2 X_3 + f_6 X_3^2,$$
$$X_0 X_2 - X_1^2 = 0,$$
$$X_0 X_3 - X_1 X_2 = 0,$$
$$X_1 X_3 - X_2^2 = 0, \qquad (1.1.5)$$

in $\mathbb{P}^4$. Here the points of (1) correspond to those of (5) with $X_0 \neq 0$ by

$$X_j = X^j, \; Y = Y. \qquad (1.1.6)$$

We denote the points on $\mathcal{C}$ by small Fraktur letters,   e.g.

$$\mathfrak{x} = (x, y). \qquad (1.1.7)$$

The point

$$\bar{\mathfrak{x}} = (x, -y) \qquad (1.1.8)$$

is the *conjugate* of $\mathfrak{x}$ (under the $\pm Y$ involution). A divisor of degree 2 of the type $\{\mathfrak{x}, \bar{\mathfrak{x}}\}$ is the intersection of $\mathcal{C}$ with $X = x$. Hence any two divisors of this type are linearly equivalent. We denote the corresponding element of Pic^2 by $\mathfrak{O}$: it is clearly the canonical class. By the Riemann-Roch Theorem, any other element $\mathfrak{A}$ of Pic^2 contains precisely one effective divisor. It is convenient, and with luck should cause no confusion, to identify $\mathfrak{A}$ with its unique effective divisor. Further, addition of $\mathfrak{O}$ in Pic identifies the jacobian $J(\mathcal{C}) = \mathrm{Pic}^0$ with Pic^2. Under this identification $0 \in J(\mathcal{C})$ goes to $\mathfrak{O}$ and the nonzero elements of $J(\mathcal{C})$ go to the effective $\mathfrak{A}$ of degree 2, $\mathfrak{A} \notin \mathfrak{O}$.

¶ If $F(X)$ is quintic it can be made sextic by using a map $X \mapsto 1/(X - \alpha)$, $Y \mapsto Y/(X - \alpha)^3$, but where now α is chosen to be any member of k which avoids the roots of $F(X)$. This is always possible when k has more than 5 elements. We shall exclude from consideration the few exceptional curves over $\mathbb{F}_3$ and $\mathbb{F}_5$ which cannot be put in the form $Y^2 = $ (sextic in X).

2   Group law. We can now describe the group law on $J(\mathcal{C})$, at least generically. Let $\mathfrak{A}$, $\mathfrak{B}$ be two effective divisors of degree 2 in general position defined over k. Then there is a unique $M(X) \in k[X]$ of degree 3 such that $Y = M(X)$ passes through the four points of $\mathfrak{A}$, $\mathfrak{B}$. The complete intersection of the cubic curve with $\mathcal{C}$ is given by

$$M(X)^2 = F(X), \quad Y = M(X). \qquad (1.2.1)$$

Hence the residual intersection is an effective divisor $\mathfrak{C}$ of degree 2, also defined over k. The divisor of poles of the function $Y - M(X)$ is at infinity, and is in the divisor class $3\mathfrak{O} \in \mathrm{Pic}^6$. Identifying elements of Pic^0 with those of Pic^2, we have thus found a $\mathfrak{C}$ such that

$$\mathfrak{A} + \mathfrak{B} + \mathfrak{C} = \mathfrak{O}. \qquad (1.2.2)$$

The divisor $\mathfrak{A} = \{\mathfrak{a}_1, \mathfrak{a}_2\}$ is rational (= defined over the ground field k) if either

(i) $\mathfrak{a}_1$ and $\mathfrak{a}_2$ are both rational

      or

(ii) they are defined over a quadratic extension of k and conjugate over k.¶

If $\mathfrak{A}$ and $\mathfrak{B}$ above are both rational, then so are $M(X)$ and $\mathfrak{C}$.

For $\mathfrak{A} = \{\mathfrak{a}_1, \mathfrak{a}_2\}$ we put $\bar{\mathfrak{A}} = \{\bar{\mathfrak{a}}_1, \bar{\mathfrak{a}}_2\}$. Clearly $\mathfrak{A} + \bar{\mathfrak{A}} = \mathfrak{O}$ in the sense of the group law, that is $\tilde{\mathfrak{A}} = -\mathfrak{A}$. It follows that $\mathfrak{A}$ is of (precise) order 2 if $\mathfrak{A} \notin \mathfrak{O}$ and $\bar{\mathfrak{A}} = \mathfrak{A}$. Hence $\mathfrak{A} = \{(\theta_1, 0), (\theta_2, 0)\}$, where θ_1, θ_2 are distinct roots of $F(X)$. Thus over $\bar{k}$ there are $\binom{6}{2} = 15$ elements of Pic^0 of order 2, though, of course, they need not be defined over k.

We reserve the further investigation of the group law until later. For a worked example, see Chapter 8, Section 1. The rest of this chapter may be omitted at a first reading. Indeed, those wishing immediately to see the more computational side of things (such as computing the group law, the torsion group and rank of the Mordell-Weil group) can at this point proceed directly to Chapter 8, followed by Chapter 11. These are designed to be fairly self-contained, provided that a few results from earlier chapters are taken on faith.

¶ There are two notions of 'conjugate' in this book: 'conjugate over k' [e.g. $(\sqrt{2}, 1 + \sqrt{2})$ and $(-\sqrt{2}, 1 - \sqrt{2})$ are conjugate over $\mathbb{Q}$] and 'conjugate under the $\pm Y$ involution' [e.g. $(\sqrt{2}, 1 + \sqrt{2})$ and $(\sqrt{2}, -1 - \sqrt{2})$]. The word 'conjugate' may be used on its own when it is clear from the context which usage applies. When $\mathfrak{x} = (x, y)$ is defined over some $k(\sqrt{d})$ we may sometimes wish to refer to both types of conjugate of $\mathfrak{x}$; we distinguish them by using $\bar{\mathfrak{x}}$ for $(x, -y)$ and $\mathfrak{x}'$ for the conjugate of $\mathfrak{x}$ over k.

3   The form is canonical. We can now confirm that every curve $\mathcal{D}$ of genus 2 defined over a perfect field k of characteristic not 2 is birationally equivalent over k to a curve (1.1).

By Riemann-Roch the differentials of the first kind¶ on $\mathcal{D}$ are a $\bar{k}$-vector space of dimension 2. By the Appendix of 'Background and Notation' there is a differential of the first kind defined over k. Its divisor of zeros $\mathfrak{H}$ is an effective divisor defined over k in the canonical class.

By Riemann-Roch again, the space of functions f with $[f] \succ -\mathfrak{H}$ has dimension 2. As before, it contains a nonconstant function X defined over k. The divisor of poles of X is $\mathfrak{H}$. By† Riemann-Roch yet again, there is a function Y defined over k which has $3\mathfrak{H}$ as divisor of poles and which is linearly independent of $1, X, X^2, X^3$. Finally,

$$\{1, X, X^2, X^3, X^4, X^5, X^6, Y, YX, YX^2, YX^3, Y^2\} \qquad (1.3.1)$$

have at worst $6\mathfrak{H}$ as divisor of poles, so are linearly dependent. After a transformation $Y \mapsto Y +$ cubic in X, this gives (1.1). The same argument shows that the transformations (1.3) are the only birational transformations which take $\mathcal{C}$ into another curve of the same form.

More generally, a hyperelliptic curve of *even* genus $g > 0$ is birationally equivalent over the ground field to a curve $Y^2 = F(X)$ with F of degree $2g + 2$; see Chevalley (1951), p. 77 [the paragraph before the enunciation of Theorem 10, but ignore from 'it can be proved' onward] or Mestre (1991b), p. 322.

4   Plane quartics. A further example of a curve of genus 2 is the plane quartic $\mathcal{D}$ with a single double point, say

$$G(U, V) = 0, \qquad (1.4.1)$$

where $G \in k[U, V]$ is of degree 4. We shall assume that the ground field k is perfect and of characteristic not 2. The double point is defined over k by Galois theory, and so can be taken to be the origin. Then

$$G(U, V) = G_2(U, V) + G_3(U, V) + G_4(U, V), \qquad (1.4.2)$$

¶ A differential is of the first kind if it has no poles.

† Alternatively, by a general theorem, the degree of the function field $k(\mathcal{D})$ over $k(X)$ is the degree of the divisor of poles of X, namely 2: and the rest is easy.

where G_j is homogeneous of degree j. On putting $V = UX$ in (2) and completing the square, (1) takes the form $\mathcal{C} : Y^2 = F(X)$ with

$$Y = 2G_4(1, X)U + G_3(1, X),$$
$$F(X) = G_3(1, X)^2 - 4G_2(1, X)G_4(1, X). \tag{1.4.3}$$

This gives a birational correspondence between $\mathcal{C}$ and $\mathcal{D}$, the double point corresponding to the divisor

$$G_2(1, X) = 0, \quad Y = G_3(1, X) \tag{1.4.4}$$

on $\mathcal{C}$.

Conversely, an effective divisor $\mathfrak{A} \not\subset \mathcal{D}$ of degree 2 on $\mathcal{C}$ is given by

$$H(X) = 0, \quad Y = M(X) \tag{1.4.5}$$

for some $H, M \in k[X]$ of degree 2 and $\leqslant 3$ respectively. Then $F(X) - M(X)^2$ is divisible by $H(X)$ and it is easy to reverse the process and to recover $\mathcal{D}$. Further, a homogeneous linear transformation of the coefficients U, V for $\mathcal{D}$ corresponds to a transformation (1.3) in the equation for $\mathcal{C}$.

Chapter 2

Construction of the jacobian

0   Introduction. In this chapter we construct the jacobian of a general curve of genus 2. A discussion of the group law is left until later. We do, however, discuss briefly the local behaviour.

1   Construction of a basis. The jacobian $J(\mathcal{C})$ is an algebraic variety whose points correspond to the elements of Pic^0. Classically it was obtained over $\mathbb{C}$ using complex-function theory, but we require an algebraic construction valid over any ground field. In the preceding chapter we identified Pic^0 with Pic^2. We thus have to take the symmetric product $\mathcal{C}^{(2)}$ of two copies of $\mathcal{C}$ and 'blow down'¶ the divisor $\mathfrak{D}$ on it consisting of the points $\{\mathfrak{x}, \bar{\mathfrak{x}}\}$.

Not every curve on a surface can be blown down: a theorem of Castelnuevo characterizes those which can. We do not need to check that his conditions are satisfied, since wiser people have already shown that the jacobian exists, at least over $\mathbb{C}$. To perform the blowing down we mimic one of the proofs of Castelnuevo's Theorem. Let

$$\mathfrak{X} = \{\mathfrak{x}, \mathfrak{u}\}, \tag{2.1.1}$$

where

$$\mathfrak{x} = (x, y), \ \mathfrak{u} = (u, v) \tag{2.1.2}$$

is a pair of generic points on $\mathcal{C}$. Then the divisor $\mathfrak{D}$ on $\mathcal{C}^{(2)}$ is given by $u = x$, $v = -y$. It turns out to be a good idea to consider functions of $\mathfrak{X}$ symmetric in $\mathfrak{x}, \mathfrak{u}$ which may have a pole of any order on $\mathfrak{D}$ and are large at worst like $x^2 u^2$ at infinity, but have no other poles. Such functions are a linear space L (say) over k. The symmetric functions of x, u of order at most 2 are included, but there are others. The function

$$\alpha_0 = \frac{y - v}{x - u} \tag{2.1.3}$$

¶ For a brief discussion of blowing up and down and of Castelnuevo's Theorem see Appendix I to this chapter.

is symmetric and small enough at infinity. At the finite poles we must have $u = x$, so $v = \pm y$. But α_0 is clearly finite on the 'diagonal' $u = x$, so its only finite pole is $\mathfrak{D}$. Hence $\alpha_0 \in L$. The square α_0^2 is too large at infinity, but by subtracting an appropriate polynomial in x, u, we obtain the element

$$\beta_0 = \frac{F_0(x, u) - 2yv}{(x - u)^2},$$

$$F_0 = 2f_0 + f_1(x + u) + 2f_2xu + f_3xu(x + u)$$
$$+ 2f_4x^2u^2 + f_5x^2u^2(x + u) + 2f_6x^3u^3 \qquad (2.1.4)$$

of L with a pole of order 2 along $\mathfrak{D}$.

Arguing along these lines we get the following 16 elements of L:

Double zero at $\mathfrak{D}$:

$$\rho = (x - u)^2. \qquad (2.1.5)$$

Regular nonzero at $\mathfrak{D}$:

$$\sigma_0 = 1, \quad \sigma_1 = x + u, \quad \sigma_2 = xu, \quad \sigma_3 = xu(x + u), \quad \sigma_4 = (xu)^2. \qquad (2.1.6)$$

Simple pole at $\mathfrak{D}$:

$$\alpha_j = \frac{u^j y - x^j v}{x - u} \quad (j = 0, 1, 2, 3). \qquad (2.1.7)$$

Double pole at $\mathfrak{D}$:

$$\beta_0 = \frac{F_0(x, u) - 2yv}{(x - u)^2}, \quad \beta_1 = \frac{F_1(x, u) - (x + u)yv}{(x - u)^2}, \quad \beta_2 = xu\beta_0, \qquad (2.1.8)$$

where

$$F_0(x, u) = 2f_0 + f_1(x + u) + 2f_2xu + f_3xu(x + u)$$
$$+ 2f_4(xu)^2 + f_5(xu)^2(x + u) + 2f_6(xu)^3,$$
$$F_1(x, u) = f_0(x + u) + 2f_1xu + f_2xu(x + u) + 2f_3(xu)^2$$
$$+ f_4(xu)^2(x + u) + 2f_5(xu)^3 + f_6(xu)^3(x + u). \qquad (2.1.9)$$

Triple pole at $\mathfrak{D}$:

$$\gamma_0 = \frac{G(x, u)y - G(u, x)v}{(x - u)^3}, \quad \gamma_1 = \frac{H(x, u)y - H(u, x)v}{(x - u)^3}, \qquad (2.1.10)$$

where

$$G(x, u) = 4f_0 + f_1(x + 3u) + f_2(2xu + 2u^2) + f_3(3xu^2 + u^3) + 4f_4xu^3$$
$$+ f_5(x^2u^3 + 3xu^4) + f_6(2x^2u^4 + 2xu^5),$$
$$H(x, u) = f_0(2x + 2u) + f_1(3xu + u^2) + 4f_2xu^2 + f_3(x^2u^2 + 3xu^3)$$
$$+ f_4(2x^2u^3 + 2xu^4) + f_5(3x^2u^4 + xu^5) + 4f_6x^2u^5. \qquad (2.1.11)$$

And, finally,
Quadruple pole at $\mathfrak{O}$:

$$\delta = \beta_0{}^2. \tag{2.1.12}$$

It is readily verified that the given functions have poles of the specified orders at $\mathfrak{O}$. It is enough to show that they are well defined at points of the diagonal $\mathfrak{X} = \{\mathfrak{x}, \mathfrak{x}\}$ not on $\mathfrak{O}$. For γ_0, for example, one checks that $G(x,u)^2 F(x) - G(u,x)^2 F(u)$ is divisible by $(x-u)^3$.

2   We have not proved that these 16 functions are a basis for L over k, but that will soon be apparent. We denote them in inverse order as $\{z_0, ..., z_{15}\}$ (so $z_0 = \delta$, $z_1 = \gamma_1$, $z_2 = \gamma_0$, $\ldots$, $z_{15} = \rho$). Anticipating its identification with the jacobian, we denote by $J(\mathcal{C})$ the projective locus¶ of $\mathbf{z} = (z_0, \ldots, z_{15})$ in $\mathbb{P}^{15}$. We show that the map from $\mathcal{C}^{(2)}$ to $J(\mathcal{C})$ is just the required blowdown: more precisely, that it is biregular outside $\mathfrak{O}$ but maps $\mathfrak{O}$ onto the single point $(1, 0, \ldots, 0)$. A divisor of the shape (1.1), (1.2) with $u \neq x$ is given by

$$\sigma_0 X^2 - \sigma_1 X + \sigma_2 = 0, \quad \sigma_0 Y = \alpha_0 X - \alpha_1; \tag{2.2.1}$$

which shows biregularity there. This extends to $u = x$, $v = y \neq 0$, on noting that there α_0, α_1 specialize to $F'(x)/2y$, $y + xF'(x)/2y$ respectively.

Finally, on dividing the projective coordinates by δ, it is easily seen that the $\mathfrak{X}$ with $u = x$, $v = -y$ (including $v = y = 0$) all go to

$$\mathbf{o} \ (\text{say}) \ = (1, 0, \ldots, 0). \tag{2.2.2}$$

Since the constructed basis behaves beautifully under the transformations induced by $X \mapsto X + \text{constant}$ and $X \mapsto X^{-1}$ (see Appendix II to this chapter), this completes the proof that $J(\mathcal{C})$ is indeed the jacobian of $\mathcal{C}$. Note that the diagonal gives an embedding of $\mathcal{C}$ in $J(\mathcal{C})$.

¶ By this is meant the projective variety defined as follows. Let $P(\mathbf{Z})$ run through the homogeneous polynomials, defined over k, in $\mathbf{Z} = (Z_0, \ldots, Z_{15})$ such that $P(\mathbf{z}) = 0$ when (x, y) and (u, v) are independent generic points of $\mathcal{C}$. The set $\mathcal{P}$ of such P is a graded ideal over $k[\mathbf{Z}]$. By Hilbert's Basissatz it has a finite basis over $k[\mathbf{Z}]$. The projective locus is the projective variety defined by the vanishing of the basis (and so of $\mathcal{P}$). In the case under discussion, a basis is given by the 72 quadratic polynomials introduced below.

We illustrate the definitions by the example

$$\mathcal{C}: \quad Y^2 = (X^2 + 1)(X^2 + 2)(X^2 + X + 1) \tag{2.2.3}$$

with $k = \mathbb{Q}$. As described in Chapter 1, the element $\mathfrak{A} = \{(i, 0), (-i, 0)\}$ of Pic^2 specifies an element of the jacobian: it is defined over $\mathbb{Q}$ [cf. discussion following (1.2.2)]. Substituting $x = i, y = 0, u = -i, v = 0$ into $(1.5), \ldots, (1.12)$ gives

$$\mathbf{z}(\mathfrak{A}) = \big(z_i(\mathfrak{A})\big) = (36, 0, 0, -6, -3, -6, 0, 0, 0, 0, 1, 0, 1, 0, 1, -4) \tag{2.2.4}$$

where $z_0, \ldots z_{15}$ are as defined at the beginning of this section. Thus, $\{(i, 0), (-i, 0)\}$ and the right hand side of (4) represent the same member of J. Similarly $\{\infty^+, \infty^+\}$ corresponds to

$$(225, -18, -60, -60, -8, 0, -8, 16, 0, 0, 16, 0, 0, 0, 0, 0), \tag{2.2.5}$$

where ∞^+ is the point at infinity for which $Y/X^3 = 1$.

3   Local behaviour. We now consider more closely the behaviour of the basis elements in the neighbourhood of $\mathfrak{O}$, say at

$$u = x + h, \quad v \approx -y, \tag{2.3.1}$$

where h is small and $y \neq 0$. It is easy to deduce from the formulae for the base elements that

$$\rho \approx h^2$$

$$\sigma_0 \approx 1 \qquad \sigma_1 \approx 2x \qquad \sigma_2 \approx x^2 \qquad \sigma_3 \approx 2x^3 \qquad \sigma_4 \approx x^4$$

$$\alpha_0 \approx \tfrac{2y}{h} \qquad \alpha_1 \approx \tfrac{2xy}{h} \qquad \alpha_2 \approx \tfrac{2x^2 y}{h} \qquad \alpha_3 \approx \tfrac{2x^3 y}{h}$$

$$\beta_0 \approx \tfrac{4y^2}{h^2} \qquad \beta_1 \approx \tfrac{4xy^2}{h^2} \qquad \beta_2 \approx \tfrac{4x^2 y^2}{h^2}$$

$$\gamma_0 \approx \tfrac{8y^3}{h^3} \qquad \gamma_1 \approx \tfrac{8xy^3}{h^3}$$

$$\delta \approx \tfrac{16y^4}{h^4} \tag{2.3.2}$$

Locally, one normalizes to $\delta = 1$. Put

$$\lambda = \frac{\gamma_0}{\delta} \approx \frac{h}{2y}, \quad \mu = \frac{\gamma_1}{\delta} \approx \frac{xh}{2y}. \tag{2.3.3}$$

Then

$$\frac{\beta_0}{\delta} \approx \lambda^2 \qquad \frac{\beta_1}{\delta} \approx \lambda\mu \qquad \frac{\beta_2}{\delta} \approx \mu^2$$

$$\frac{\alpha_0}{\delta} \approx \lambda^3 \qquad \frac{\alpha_1}{\delta} \approx \lambda^2\mu \qquad \frac{\alpha_2}{\delta} \approx \lambda\mu^2 \qquad \frac{\alpha_3}{\delta} \approx \mu^3$$

$$\frac{\sigma_0}{\delta} \approx \lambda^4 \quad \frac{\sigma_1}{\delta} \approx 2\lambda^3\mu \quad \frac{\sigma_2}{\delta} \approx \lambda^2\mu^2 \quad \frac{\sigma_3}{\delta} \approx 2\lambda\mu^3 \quad \frac{\sigma_4}{\delta} \approx \mu^4 \quad (2.3.4)$$

and

$$\frac{\rho}{\delta} \approx \frac{h^6}{16y^4} = \frac{h^6 F(x)}{16y^6} \approx 4 \sum_{j=0}^{6} f_j \lambda^{6-j} \mu^j \qquad (2.3.5)$$

Using the local behaviour, one can determine a basis for the quadratic relations between the base elements: for example $\alpha_0{}^2$ and $\beta_0\sigma_0$ have the same dominant term λ^6, so their difference must be expressible in terms of 'smaller' products. The reader can verify that in fact

$$z_9^2 - z_5 z_{14} - f_2 z_{14}^2 - f_3 z_{14} z_{13} - f_4 z_{13}^2 - 3 f_5 z_{13} z_{12}$$
$$- f_5 z_{13} z_{15} - f_6 z_{14} z_{10} - 6 f_6 z_{12} z_{15} - 8 f_6 z_{12}^2 - f_6 z_{15}^2 = 0, \qquad (2.3.6)$$

where $z_0, \ldots, z_{15}$ are as defined at the beginning of Section 2.

It turns out that the set of quadratic relations has dimension 72 over k. Hence $L^{\otimes 2}$ has dimension $\binom{16+1}{2} - 72 = 64$ in accordance with the classical theory. The jacobian is given by the quadratic relations, in conformity with a theorem of Mumford (1966, I) about the defining equations of abelian varieties. We have made a basis for the quadratic relations (and so a set of defining equations for the jacobian) available by anonymous ftp, as explained in Appendix II at the end of the book. Those familiar with the classical theory will also note the similarity between the form of the *local parameters* λ, μ, and the fact that dX/Y and XdX/Y are a basis of the differentials of the first kind on $\mathcal{C}$. For details see Flynn (1990a).¶

4   Formal group law. We leave the derivation of the group law on the jacobian until the end of Chapter 3. The corresponding formal group is discussed fully in Chapter 7, but we introduce it briefly here. This section may be omitted in a first reading.

¶ The notation in Flynn's earlier papers differs from that of these prolegomena. Apart from the use of different symbols, the main difference is that they have $x^2 + u^2$ as a basis element instead of $\rho = (x - u)^2$. The files available by ftp use the notation introduced here.

Note first that the base elements, normalized to $\delta = 1$, may be expressed to arbitrary precision as formal power series in λ, μ by the techniques of the previous section.

Let $\mathfrak{A}, \mathfrak{B}$ be generic divisors on $\mathcal{C}$. Then $\lambda(\mathfrak{A} + \mathfrak{B})$ and $\mu(\mathfrak{A} + \mathfrak{B})$ are expressible as formal power series in the parameters of $\mathfrak{A}, \mathfrak{B}$. It is fairly straightforward to determine the power series to arbitrary accuracy by determining the coefficients by successive approximation using the formulae of Chapter 1, Section 3 or the biquadratic forms of Chapter 3, Section 4. To get the process going, we may need to know that the formal law is approximately linear, and this we do here. The technique, introducing a large constant factor M in an expression for Y, turns out to be of general application in local questions at $\mathbf{o}$.

Let $\mathfrak{r}_j = (x_j, y_j)$ $(j = 1, 2, 3)$ be three points on $\mathcal{C}$ with distinct x_j and $y_j \neq 0$. At the six intersections of $\mathcal{C}$ with

$$Y = M(X - x_1)(X - x_2)(X - x_3), \qquad (2.4.1)$$

we have

$$M^2(X - x_1)^2(X - x_2)^2(X - x_3)^2 = Y^2 = F(X). \qquad (2.4.2)$$

Suppose that M is large. The intersections occur in pairs, two near each $\mathfrak{r}_j$. Call them $(x_j^\pm, y_j^\pm)$ with $x_j^\pm \approx x_j$, $y_j^\pm \approx \pm y_j$. From (2) we have

$$M(x_1^\pm - x_1) \approx \frac{\pm y_1}{(x_1 - x_2)(x_1 - x_3)}, \qquad (2.4.3)$$

for $j = 1$, and similarly for $j = 2, 3$. It follows that the values λ_1, μ_1 of the parameters λ, μ for $\mathfrak{X}_1 = \{(x_1^+, y_1^+), (x_1^-, y_1^-)\}$ satisfy

$$M\lambda_1 \approx \frac{4}{(x_1 - x_2)(x_1 - x_3)}, \quad M\mu_1 \approx \frac{4x_1}{(x_1 - x_2)(x_1 - x_3)}. \qquad (2.4.4)$$

Then in an obvious notation

$$M\sum_j \lambda_j \approx 0, \quad M\sum_j \mu_j \approx 0. \qquad (2.4.5)$$

For indeterminate values of the parameters f_j, one checks by induction that the coefficients in the formal law are in $\mathbb{Z}[f_0, \ldots, f_6]$. If the f_j are given values as integers in a non-archimedean valued field, it follows that the power series converge for values of the parameters in the valuation ideal.

Appendix I

Blowing up and blowing down

We briefly recall the theory. Let P be a point on the nonsingular projective surface S. Then there is a nonsingular projective surface S^*, a curve $\mathcal{P}$ on it, and a birational correspondence between S and S^* which is biregular outside P and $\mathcal{P}$, but which makes P correspond to the whole of $\mathcal{P}$. More precisely, the points on $\mathcal{P}$ correspond to the directions through P.

We illustrate the process in the simplest case when S is the projective plane, say with coordinates (U_1, U_2, U_3) and $P = (1, 0, 0)$. Let $\mathcal{L}$ be the projective line, say with coordinates (V_1, V_2). Consider the product variety whose coordinates are

$$X_1 = V_1 U_1, \quad X_2 = V_1 U_2, \quad X_3 = V_1 U_3,$$
$$X_4 = V_2 U_1, \quad X_5 = V_2 U_2, \quad X_6 = V_2 U_3. \tag{2.I.1}$$

This is the variety in $\mathbb{P}^5$ given by the equations

$$X_1 X_5 = X_2 X_4, \ X_1 X_6 = X_3 X_4, \ X_2 X_6 = X_3 X_5. \tag{2.I.2}$$

Now impose the condition that the direction from P to (U_1, U_2, U_3) is just (V_1, V_2), that is $V_2 U_3 = V_1 U_2$, or $X_6 = X_2$. The result is the variety in $\mathbb{P}^4$ with coordinates X_1, X_2, X_3, X_4, X_5 satisfying

$$X_1 X_5 = X_2 X_4, \ X_1 X_2 = X_3 X_4, \ X_2^2 = X_3 X_5. \tag{2.I.3}$$

This is easily seen to be the required S^*. The curve $\mathcal{P}$ is $X_2 = X_3 = X_5 = 0$.

Blowing up is a much-used tool: for example the general cubic surface is obtained by blowing up six points in the projective plane simultaneously. [For applications in a number-theoretical context, see H.P.F. Swinnerton-Dyer (1971).]

The reverse process to blowing up is blowing down. A curve on a surface can only very exceptionally be blown down to a point. The conditions for this to be possible are expressed in a famous theorem of Castelnuevo, which we need not discuss further here.

Appendix II

Change of coordinates on $\mathcal{C}$

We saw in Chapter 1 that the curve

$$\mathcal{C}: \quad Y^2 = F(X) = \sum_{j=0}^{6} f_j X^j \tag{2.II.1}$$

is unique in its birational equivalence class only up to a fractional linear transformation of X and an associated transformation of Y [see (1.3) and its footnote]. The jacobian belongs to the equivalence class and not to one particular curve in it.

In this appendix we study how the basis we have constructed above for the jacobian behaves under the change from one curve in the class to another.¶ For example, the set of all quadratic relations satisfied by the basis elements must be preserved, which gives a very severe restriction on their possible shape. This was a useful technique in exploratory work, but now we can do without it.

Denote the new variables by X^*, Y^* and the corresponding equation by

$$\mathcal{C}^*: \quad Y^{*2} = F^*(X^*) = \sum f_j^* X^{*j}. \tag{2.II.2}$$

In general we use $*$ to indicate expressions defined in terms of (2). The general transformation from (1) to (2) can be broken into a succession of simpler transformations, which we treat separately.

First, consider
$$X^* = X, \quad Y^* = tY, \tag{2.II.3}$$

where t is a parameter. It is easily seen that

$$\begin{aligned}
f_j^* &= t^2 f_j, \\
\rho^* &= \rho, \\
\sigma_j^* &= \sigma_j, \\
\alpha_j^* &= t\alpha_j, \\
\beta_j^* &= t^2 \beta_j, \\
\gamma_j^* &= t^3 \gamma_j, \\
\delta^* &= t^4 \delta.
\end{aligned} \tag{2.II.4}$$

¶ For a set of invariants of $\mathcal{C}$ under these transformations, see Igusa (1960), Mestre (1991b) and Liu (1994).

It follows that any minimal polynomial relation between the f_j and the basis elements is homogeneous in t under the substitution (4).

Next, consider

$$X^* = tX, \quad Y^* = Y. \tag{2.II.5}$$

Then

$$
\begin{aligned}
f_j^* &= t^{-j} f_j, \\
\rho^* &= t^2 \rho, \\
\sigma_j^* &= t^j \sigma_j, \\
\alpha_j^* &= t^{j-1} \alpha_j, \\
\beta_j^* &= t^{j-2} \beta_j, \\
\gamma_j^* &= t^{j-3} \gamma_j, \\
\delta^* &= t^{-4} \delta.
\end{aligned}
\tag{2.II.6}
$$

This again gives a homogeneity condition on a minimum polynomial relation between the f_j and the basis.

The next transformation is an involution, so it is convenient to introduce a different notation.

$$\hat{X} = X^{-1}, \quad \hat{Y} = X^{-3} Y. \tag{2.II.7}$$

The basis elements are a homogeneous set and so can be multiplied throughout by a common multiplier. We take this multiplier to be $(xu)^2$ and then have

$$
\begin{aligned}
\hat{f}_j &= f_{6-j}, \\
\hat{\rho} &= \rho, \\
\hat{\sigma}_j &= \sigma_{4-j}, \\
\hat{\alpha}_j &= \alpha_{3-j}, \\
\hat{\beta}_j &= \beta_{2-j}, \\
\hat{\gamma}_j &= \gamma_{1-j}, \\
\hat{\delta} &= \delta.
\end{aligned}
\tag{2.II.8}
$$

A minimal relation is either symmetric or anti-symmetric under the involution ($\hat{}$).

There remains the most difficult case

$$X^* = X + t, \quad Y^* = Y. \tag{2.II.9}$$

Let **R** be the ring generated over $\mathbb{Q}$ by the coefficients f_j and the 16 basis elements $\delta, \ldots, \rho$. The transformation $*$ induced by (9) extends to a map,¶ also denoted by $*$, of $\mathbf{R}$ into $\mathbf{R}[t]$.

For $\xi \in \mathbf{R}$ denote by $\mathrm{D}\xi$ the coefficient of t in ξ^*. Then D is a differentiation on $\mathbf{R}$ and it is easily seen that

$$\xi^* = (\exp t\mathrm{D})(\xi) \qquad (2.\text{II}.10)$$

in the sense of the 19th century algebra of operators. In particular, $\mathrm{D}\xi = 0$ if and only if $\xi^* = \xi$. One computes fairly readily that

$$\mathrm{D}f_6 = 0, \quad \mathrm{D}f_j = -(j+1)f_{j+1} \ (0 \leqslant j \leqslant 5); \qquad (2.\text{II}.11a)$$

$$\mathrm{D}\rho = 0; \qquad (2.\text{II}.11b)$$

$$\mathrm{D}\sigma_0 = 0, \ \mathrm{D}\sigma_1 = 2\sigma_0, \ \mathrm{D}\sigma_2 = \sigma_1, \ \mathrm{D}\sigma_3 = 6\sigma_2 + \rho, \ \mathrm{D}\sigma_4 = 2\sigma_3; \qquad (2.\text{II}.11c)$$

$$\mathrm{D}\alpha_0 = 0, \ \mathrm{D}\alpha_1 = \alpha_0, \ \mathrm{D}\alpha_2 = 2\alpha_1, \ \mathrm{D}\alpha_3 = 3\alpha_2; \qquad (2.\text{II}.11d)$$

$$\mathrm{D}\beta_0 = f_3\sigma_0 + 2f_5\sigma_2, \ \mathrm{D}\beta_1 = f_2\sigma_0 + 2f_4\sigma_2 + 3f_6\sigma_4 + \beta_0,$$
$$\mathrm{D}\beta_2 = f_1\sigma_0 + 2f_3\sigma_2 + 3f_5\sigma_4 + 2\beta_1; \qquad (2.\text{II}.11e)$$

$$\mathrm{D}\gamma_0 = 3f_5\alpha_2 + 2f_6\alpha_3, \ \mathrm{D}\gamma_1 = 2f_3\alpha_1 + 2f_4\alpha_2 + f_5\alpha_3 + \gamma_0; \qquad (2.\text{II}.11f)$$

$$\mathrm{D}\delta = 2f_3\beta_0 + 4f_5\beta_2. \qquad (2.\text{II}.11g)$$

One can replace the β_j, γ_j, δ in the basis for the jacobian by new quantities which behave better under ($\hat{\ }$) and D. To avoid denominators write, temporarily,

$$f_j = \binom{6}{j} g_j, \qquad (2.\text{II}.12)$$

so

$$\mathrm{D}g_j = -(6-j)g_{j+1} \ (0 \leqslant j \leqslant 5), \quad \mathrm{D}g_6 = 0. \qquad (2.\text{II}.13)$$

Put

$$\tilde{\beta}_0 = \beta_0 + 6g_4\sigma_2 + 2g_3\sigma_1 + 6g_2\sigma_0,$$
$$\tilde{\beta}_1 = \beta_1 + 3g_5\sigma_4 + 3g_4\sigma_3 + 14g_3\sigma_2 + 3g_2\sigma_1 + 3g_1\sigma_0 + g_3\rho,$$
$$\tilde{\beta}_2 = \beta_2 + 6g_4\sigma_4 + 2g_3\sigma_3 + 6g_2\sigma_2;$$
$$\tilde{\gamma}_0 = \gamma_0 + 2g_5\alpha_3 + 12g_4\alpha_2 + 8g_3\alpha_1 + 2g_2\alpha_0, \qquad (2.\text{II}.14)$$
$$\tilde{\gamma}_1 = \gamma_1 + 2g_4\alpha_3 + 8g_3\alpha_2 + 12g_2\alpha_1 + 2g_1\alpha_0;$$
$$\tilde{\delta} = \tilde{\beta}_0{}^2.$$

¶ More precisely, let $\mathfrak{X} = \{(x,y),(u,v)\}$ be a divisor on (1) and let (x^*,y^*), (u^*,v^*) be the points on (2) corresponding to (x,y), (u,v). Let ζ be one of the 16 basis elements defined in Section 1 as a function of x,y,u,v and the f_j. Then ζ^* is the same function of x^*,y^*,u^*,v^* and the f_j^*.

Then
$$\widehat{\tilde{\beta}}_j = \tilde{\beta}_{2-j},$$
$$\widehat{\tilde{\gamma}}_j = \tilde{\gamma}_{1-j}, \qquad (2.\text{II}.15)$$
$$\widehat{\tilde{\delta}} = \tilde{\delta}$$

and
$$\text{D}\tilde{\beta}_2 = 2\tilde{\beta}_1, \ \text{D}\tilde{\beta}_1 = \tilde{\beta}_0, \ \text{D}\tilde{\beta}_0 = 0;$$
$$\text{D}\tilde{\gamma}_1 = \tilde{\gamma}_0, \ \text{D}\tilde{\gamma}_0 = 0; \qquad (2.\text{II}.16)$$
$$\text{D}\tilde{\delta} = 0.$$

Further,
$$\tilde{\beta}_2 = (xu)\beta_0, \ 2\tilde{\beta}_1 = (x+u)\tilde{\beta}_0. \qquad (2.\text{II}.17)$$

The local parameters
$$\tilde{\lambda} = \tilde{\gamma}_0/\tilde{\delta}, \ \tilde{\mu} = \tilde{\gamma}_1/\tilde{\delta} \qquad (2.\text{II}.18)$$

at **o** satisfy
$$\widehat{\tilde{\mu}} = \tilde{\lambda}, \ \widehat{\tilde{\lambda}} = \tilde{\mu}, \ \text{D}\tilde{\mu} = \tilde{\lambda}, \ \text{D}\tilde{\lambda} = 0. \qquad (2.\text{II}.19)$$

Defining for completeness $\tilde{\rho} = \rho$, $\tilde{\sigma}_j = \sigma_j$, $\tilde{\alpha}_j = \alpha_j$, we see that the 'twiddles' ($\tilde{\ }$) basis is much better behaved under transformations of X, Y. One might expect this to be reflected in a greater simplicity in the shape of, say, the addition formulae. This does not appear to be the case, and as the expressions for the twiddles basis in terms of x, y, u, v are more complicated, we do not use it further.

Chapter 3

The Kummer surface

0   Introduction. The jacobian $J(\mathcal{C})$ of a curve $\mathcal{C}$ of genus 2, which we have constructed in Chapter 2, is a rather large object which is difficult to manipulate, even with the resources of computer algebra. In this chapter we shall see that much of its information is already contained in a much simpler object, the associated Kummer surface $\mathcal{K} = \mathcal{K}(\mathcal{C})$. This is a quartic surface in $\mathbb{P}^3$.

Kummer surfaces over $\mathbb{C}$ and $\mathbb{R}$ were discovered by Kummer (1864a,b; 1866) in the context of a problem in the diffraction of light. They are remarkable for having the maximum number, 16, of point singularities for a quartic surface in $\mathbb{P}^3$ which has no singular curves. The relationship with abelian varieties, and so to the present circle of ideas, emerged only later.¶ Kummer surfaces were extensively studied: Hudson (1905)† gives a readable account which has recently been republished. Models of different surfaces were produced commercially and are to be found in some of the older universities:‡ there are illustrations in Fischer (1986).

From our standpoint, the Kummer surface $\mathcal{K}$ belonging to a curve of genus 2 is the projective locus in $\mathbb{P}^3$ of four of the 16 basis elements constructed in Chapter 2. We shall use the notation

$$\xi_1 = \sigma_0,\ \xi_2 = \sigma_1,\ \xi_3 = \sigma_2,\ \xi_4 = \beta_0. \tag{3.0.1}$$

Since the ξ_j are a subset of the basis elements z_j which define the jacobian, there is an obvious map from the jacobian to the Kummer. As we shall see below, two points of the jacobian go into the same point of the Kummer precisely when they are related by the involution $\mathfrak{X} \mapsto -\mathfrak{X}$. The fixed points

---

¶ Cayley (1877) and Borchardt (1877) independently. The papers are in the same volume of *Crelle*.

† Ronald William Henry Turnbull Hudson (1876–1904) was one of the select band of mathematicians who died in a climbing accident at an early age. The book was published posthumously.

‡ What is almost certainly the model in the frontispiece of Hudson (1905) is in the collection of the Department of Pure Mathematics and Mathematical Statistics at Cambridge.

under the involution are the 16 points of order 2 on the jacobian, and these map into the 16 singularities of the Kummer (the *nodes*). The jacobian is a double cover of the Kummer ramified precisely at the nodes.

The reader familiar with the theory of elliptic curves will note analogies: the analogue of the jacobian and the Kummer are the elliptic curve itself ($Y^2 = X^3 + AX + B$, say) and the line with coordinate X.

In this chapter we describe the Kummer and some of its more basic properties. These are used to give an explicit treatment of the group law on the jacobian. In a final section we discuss when a surface with the geometric properties of a Kummer surface actually arises from a curve.

We do not reproduce some of the more complicated formulae. They are available by ftp as explained in Appendix II. Appendix I gives a program for the algebra package MAPLE which verifies much of the work: this is also available by ftp. For an earlier treatment in a rather different notation, see Flynn (1993).

1   Construction of the surface. Nodes.

We shall use the notation of Chapter 2, some of which we briefly recall. A general ('generic') divisor of degree 2 on

$$\mathcal{C}: \ Y^2 = F(X) = \sum_{j=0}^{6} f_j X^j \tag{3.1.1}$$

is denoted by

$$\mathfrak{X} = \{(x,y),(u,v)\}. \tag{3.1.2}$$

We are primarily concerned with four of the 16 basis elements of Chapter 2, which we now denote $\xi_j = \xi_j(\mathfrak{X})$ as follows:

$$\xi_1 = \sigma_0 = 1, \ \xi_2 = \sigma_1 = (x+u), \ \xi_3 = \sigma_2 = xu, \ \xi_4 = \beta_0, \tag{3.1.3}$$

where

$$\beta_0 = \frac{F_0(x,u) - 2yv}{(x-u)^2}. \tag{3.1.4}$$

Here

$$F_0(x,u) = 2f_0 + f_1(x+u) + 2f_2(xu) + f_3(x+u)(xu) \\ + 2f_4(xu)^2 + f_5(x+u)(xu)^2 + 2f_6(xu)^3. \tag{3.1.5}$$

Note that $F_0(x,u)^2 - 4F(x)F(u)$ is divisible by $(x-u)$, and so, since it is symmetric, by $(x-u)^2$. Hence β_0 satisfies the quadratic equation

$$A{\beta_0}^2 + B\beta_0 + C = 0, \tag{3.1.6}$$

where

$$A = (x - u)^2$$
$$B = -2F_0(x, u)$$
$$C = \frac{F_0(x, u)^2 - 4F(x)F(u)}{(x - u)^2}$$

(3.1.7)

are symmetric polynomials in x, u of degree at most 2, 3, 4 in the variables separately. Hence the locus of (3) in $\mathbb{P}^3$ is given by

$$K(\xi_1, \ldots, \xi_4) = K_2{\xi_4}^2 + K_1\xi_4 + K_0 = 0,$$

(3.1.8)

where

$$
\begin{aligned}
K_2 &= {\xi_2}^2 - 4\xi_1\xi_3, \\
K_1 &= -2\{2f_0{\xi_1}^3 + f_1{\xi_1}^2\xi_2 + 2f_2{\xi_1}^2\xi_3 + f_3\xi_1\xi_2\xi_3 \\
&\quad + 2f_4\xi_1{\xi_3}^2 + f_5\xi_2{\xi_3}^2 + 2f_6{\xi_3}^3\}, \\
K_0 &= (f_1^2 - 4f_0f_2)\,{\xi_1}^4 - 4f_0f_3{\xi_1}^3\xi_2 - 2f_1f_3{\xi_1}^3\xi_3 - 4f_0f_4{\xi_1}^2{\xi_2}^2 \\
&\quad + 4(f_0f_5 - f_1f_4)\,{\xi_1}^2\xi_2\xi_3 + (f_3^2 + 2f_1f_5 - 4f_2f_4 - 4f_0f_6)\,{\xi_1}^2{\xi_3}^2 \\
&\quad - 4f_0f_5\xi_1{\xi_2}^3 + 4(2f_0f_6 - f_1f_5)\,\xi_1{\xi_2}^2\xi_3 + 4(f_1f_6 - f_2f_5)\,\xi_1\xi_2{\xi_3}^2 \\
&\quad - 2f_3f_5\xi_1{\xi_3}^3 - 4f_0f_6{\xi_2}^4 - 4f_1f_6{\xi_2}^3\xi_3 - 4f_2f_6{\xi_2}^2{\xi_3}^2 \\
&\quad - 4f_3f_6\xi_2{\xi_3}^3 + (f_5^2 - 4f_4f_6)\,{\xi_3}^4.
\end{aligned}
$$

(3.1.9)

The *Kummer surface* (or *Kummer*) is the surface $K(\xi_1, \xi_2, \xi_3, \xi_4) = 0$.

There are minor subtleties in the map from the jacobian $J(\mathcal{C})$ to the Kummer $\mathcal{K}$ which will be analysed in Section 8, but the general picture is clear. Recall the behaviour of the basis $z_0, \ldots, z_{15} = \delta, \ldots, \rho$ in Chapter 2 under the involution

$$\mathfrak{X} = \{(x, y), (u, v)\} \longmapsto -\mathfrak{X} = \{(x, -y), (u, -v)\}.$$

(3.1.10)

The α_j and γ_j are odd but the remaining basis elements are even. Further, the ten even basis elements are homogeneous quadratic in the ξ_j given by (3). They are linearly independent, and so span the 10-dimensional space of homogeneous quadratic forms in the ξ_j. Hence the map from $J(\mathcal{C})$ to $\mathcal{K}$ is biregular except at the points belonging to divisor classes $\mathfrak{X}$ with

$$\mathfrak{X} = -\mathfrak{X},$$

(3.1.11)

i.e. $2\mathfrak{X} = \mathfrak{O}$. One of these points is $\mathfrak{O}$ itself. Its image in $\mathcal{K}$ is $\mathbf{o} = (0, 0, 0, 1)$.

In the neighbourhood of **o** the Kummer behaves like the cone $K_2 = 0$. In the accepted terminology, **o** is a *node*. The other $\mathfrak{X}$ satisfying (11) are the $\{(x,0),(u,0)\} \notin \mathfrak{D}$. These have $F(x) = F(u) = 0$, $x \neq u$. Let θ_j $(j = 1,\ldots,6)$ be the roots of $F(X)$ in $\bar{k}$. Then there are just $\binom{6}{2} = 15$ divisors of order (precisely) 2, namely the $\{(\theta_i,0),(\theta_j,0)\}$ with, say, $i < j$. We denote the corresponding point on the Kummer by $N_{ij} = N_{ji}$. By transport of structure the behaviour of the Kummer in the neighbourhood of N_{ij} is similar to that in the neighbourhood of **o**, which we also denote by N_0.

Hence the Kummer has 16 nodes in all, which, as we remarked above, is the maximum possible for a quartic surface. We first compute their coordinates. Put temporarily

$$F(X) = (X - \theta_i)(X - \theta_j)H(X), \quad H(X) = \sum_{n=0}^{4} h_n X^n. \qquad (3.1.12)$$

Then it is readily computed that

$$\begin{aligned} \beta_0(\{(\theta_i,0),(\theta_j,0)\}) &= \beta_0(i,j) \text{ (say)} \\ &= -h_0 - h_2(\theta_i\theta_j) - h_4(\theta_i\theta_j)^2. \end{aligned} \qquad (3.1.13)$$

Hence N_{ij} has coordinates

$$N_{ij}: \quad (1, \theta_i + \theta_j, \theta_i\theta_j, \beta_0(i,j)). \qquad (3.1.14)$$

2   Addition of 2-division points. In passing from the jacobian to the Kummer we have lost the group structure, but traces of it remain. In particular, addition of an element of order 2 remains meaningful, We shall show that it is given by a linear transformation of the ambient space. It is defined over the field of definition of the element of order 2.

The construction of the linear transformation is straightforward. We give sufficient detail to enable the reader to check (preferably with an algebra package!). We have listed a suggested MAPLE program which does this, amongst other things, in Appendix I [Part 2 of the program Kummer.pro]. Let $\mathfrak{X} = \{(x,y),(u,v)\}$ be a generic divisor and $\mathfrak{A} = \{(\theta_1,0),(\theta_2,0)\}$. Put

$$G(X) = (X - \theta_1)(X - \theta_2), \quad F(X) = G(X)H(X). \qquad (3.2.1)$$

We have to find the point on the Kummer belonging to the residual intersection of $Y = M(X)$ with $\mathcal{C}$, where the cubic $M(X)$ is chosen so that it passes through $\mathfrak{X}$ and $\mathfrak{A}$. Hence

$$M(X) = G(X)M_0(X), \qquad (3.2.2)$$

where

$$M_0(X) = \frac{X - x}{u - x} \frac{v}{G(u)} + \frac{X - u}{x - u} \frac{y}{G(x)}. \qquad (3.2.3)$$

In $M_0(X)^2$ substitute $y^2 = F(x)$, $v^2 = F(u)$ and eliminate yv by the definition of $\beta_0 = \beta_0(\mathfrak{X})$. Then

$$G(x)G(u)M_0(X)^2 = (X - x)(X - u)\beta_0 + L, \qquad (3.2.4)$$

where

$$L = \frac{(X - x)^2 G(x)H(u) + (X - u)^2 G(u)H(x) - (X - x)(X - u)F_0(x, u)}{(x - u)^2} \qquad (3.2.5)$$

is actually a polynomial in X, x, u, symmetric in x, u. On substituting $Y = M(X)$ in the equation for $\mathcal{C}$ we find that the X-coordinates of the residual intersection satisfy

$$\beta_0(X - x)(X - u) + N = 0, \qquad (3.2.6)$$

where

$$N = \frac{G(X)L - H(X)G(x)G(u)}{(X - x)(X - u)} \qquad (3.2.7)$$

is actually a polynomial in X, x, u quadratic in X and linear and symmetric in x, u. We can therefore write (6) as

$$\xi_0{}^* X^2 - \xi_1{}^* X + \xi_2{}^* = 0, \qquad (3.2.8)$$

where the $\xi_j{}^*$ are linear forms in the ξ_j (given by (1.3)). We have thus found (the ratios of) three of the coordinates of the residual intersection. Write

$$\xi_i{}^* = \sum_{j=1}^{4} w_{ij}\xi_j \qquad (i = 1, 2, 3) \qquad (3.2.9)$$

where the w_{ij} are given by¶

$$\begin{bmatrix} g_2^2 h_0 + g_0 g_2 h_2 - g_0^2 h_4 & g_0 g_2 h_3 - g_0 g_1 h_4 & g_1 g_2 h_3 - g_1^2 h_4 + 2g_0 g_2 h_4 & g_2 \\ -g_0 g_2 h_1 - g_0 g_1 h_2 + g_0^2 h_3 & g_2^2 h_0 - g_0 g_2 h_2 + g_0^2 h_4 & g_2^2 h_1 - g_1 g_2 h_2 - g_0 g_2 h_3 & -g_1 \\ -g_1^2 h_0 + 2g_0 g_2 h_0 + g_0 g_1 h_1 & -g_1 g_2 h_0 + g_0 g_2 h_1 & -g_2^2 h_0 + g_0 g_2 h_2 + g_0^2 h_4 & g_0 \end{bmatrix}$$

$$(3.2.10a)$$

¶ Here g_j, h_j is the coefficient of X^j in $G(X)$, $H(X)$ respectively, so $g_2 = 1$, $h_4 = f_6$.

But we are adding a point of order 2 and so (9) holds up to a common proportional factor with the starred and unstarred coordinates interchanged. Otherwise expressed, the w_{ij} form part of a 4×4 matrix $\mathbf{W}$ such that $\mathbf{W}^2$ is a scalar multiple of the unit matrix $\mathbf{I}$. This gives the remaining line, namely

$$
\begin{aligned}
w_{41} = \ & -g_1 g_2^2 h_0 h_1 + g_1^2 g_2 h_0 h_2 + g_0 g_2^2 h_1^2 - 4 g_0 g_2^2 h_0 h_2 \\
& - g_0 g_1 g_2 h_1 h_2 + g_0 g_1 g_2 h_0 h_3 - g_0^2 g_2 h_1 h_3, \\
w_{42} = \ & g_1^2 g_2 h_0 h_3 - g_1^3 h_0 h_4 - 2 g_0 g_2^2 h_0 h_3 - g_0 g_1 g_2 h_1 h_3 \\
& + 4 g_0 g_1 g_2 h_0 h_4 + g_0 g_1^2 h_1 h_4 - 2 g_0^2 g_2 h_1 h_4, \\
w_{43} = \ & -g_0 g_2^2 h_1 h_3 - g_0 g_1 g_2 h_2 h_3 + g_0 g_1 g_2 h_1 h_4 + g_0 g_1^2 h_2 h_4 \\
& + g_0^2 g_2 h_3^2 - 4 g_0^2 g_2 h_2 h_4 - g_0^2 g_1 h_3 h_4, \\
w_{44} = \ & -g_2^2 h_0 - g_0 g_2 h_2 - g_0^2 h_4.
\end{aligned}
\qquad (3.2.10b)
$$

The constant c such that $\mathbf{W}^2 = c\mathbf{I}$ is effectively the resultant of $G(X)$ and $H(X)$. More precisely, if θ_1, θ_2 are the roots of $G(X)$ we have

$$
c = H(\theta_1) H(\theta_2) = f_6^2 \prod_{j=3}^{6} (\theta_1 - \theta_j)(\theta_2 - \theta_j).
\qquad (3.2.11)
$$

3   Commutativity properties. Let $\mathbf{W}[ij]$ for $i \neq j$ be the matrix constructed as above for the divisor $\{(\theta_i, 0), (\theta_j, 0)\}$ which, just for the moment, we denote by the same symbol N_{ij} as the corresponding node. Together with $N_0 = 0$ they form a group of order 16 and exponent 2. Typical relations are

$$
2N_{ij} = N_0 = 0, \qquad N_{12} + N_{13} = N_{23}, \qquad N_{12} + N_{34} = N_{56}, \qquad (3.3.1)
$$

The transformations induced by the $\mathbf{W}[ij]$ in $\mathbb{P}^3$ must represent the same group. The $\mathbf{W}[ij]$, however, do not themselves commute. A straightforward, if tedious, calculation shows that

$$
\begin{aligned}
\mathbf{W}[12]\mathbf{W}[13] &= c\mathbf{W}[23], \\
c &= f_6(\theta_1 - \theta_4)(\theta_1 - \theta_5)(\theta_1 - \theta_6)(\theta_2 - \theta_3),
\end{aligned}
\qquad (3.3.2)
$$

and

$$
\begin{aligned}
\mathbf{W}[12]\mathbf{W}[34] &= d\mathbf{W}[56], \\
d &= -f_6(\theta_1 - \theta_3)(\theta_1 - \theta_4)(\theta_2 - \theta_3)(\theta_2 - \theta_4).
\end{aligned}
\qquad (3.3.3)
$$

This implies that
$$\mathbf{W}[ij]\mathbf{W}[lm] = \epsilon\mathbf{W}[lm]\mathbf{W}[ij], \tag{3.3.4}$$

where
$$\epsilon = \begin{cases} -1 & \text{if card}(\{i,j\} \cap \{l,m\}) = 1, \\ +1 & \text{otherwise.} \end{cases} \tag{3.3.5}$$

Cognoscenti will recognize the *Weil pairing*.

4   The biquadratic forms. For general divisors $\mathfrak{A}$, $\mathfrak{B}$ we cannot determine the $\xi_j(\mathfrak{A} + \mathfrak{B})$ from the values of the $\xi_j(\mathfrak{A})$, $\xi_j(\mathfrak{B})$ alone, since the latter do not distinguish between $\pm\mathfrak{A}$, $\pm\mathfrak{B}$, and so not between $\mathfrak{A} \pm \mathfrak{B}$. However the $\xi_i(\mathfrak{A} + \mathfrak{B})\,\xi_j(\mathfrak{A} - \mathfrak{B}) + \xi_i(\mathfrak{A} - \mathfrak{B})\,\xi_j(\mathfrak{A} + \mathfrak{B})$ are determined. We have

THEOREM 3.4.1. *There are polynomials B_{ij} biquadratic in the $\xi_j(\mathfrak{A})$, $\xi_j(\mathfrak{B})$ such that projectively*

$$\left(\xi_i(\mathfrak{A} + \mathfrak{B})\,\xi_j(\mathfrak{A} - \mathfrak{B}) + \xi_i(\mathfrak{A} - \mathfrak{B})\,\xi_j(\mathfrak{A} + \mathfrak{B})\right) = \left(2B_{ij}(\mathfrak{A},\mathfrak{B})\right). \tag{3.4.1}$$

If the coordinates are normalized to $\xi_1(\mathfrak{A}) = \xi_1(\mathfrak{B}) = \xi_1(\mathfrak{A} \pm \mathfrak{B}) = 1$, then the coefficients in the B_{ij} are in $\mathbb{Z}[f_0, \ldots, f_6]$.

For the analogue for elliptic curves cf. Chapter 17, Lemma 4 and Chapter 24, Lemma 3 of Cassels (1991).

We already know the values of (1) when $\mathfrak{B} = \mathfrak{D}$ is of order 2, since

$$\left(B_{ij}(\mathfrak{A},\mathfrak{D})\right) = \left(\xi_i(\mathfrak{A} + \mathfrak{D})\,\xi_j(\mathfrak{A} + \mathfrak{D})\right), \tag{3.4.2}$$

where we saw in the last section that the two factors on the right are linear in the $\xi_j(\mathfrak{A})$ for generic $\mathfrak{A}$. To establish the required formulae for (1) it is enough to consider the general case when the f_j are independent transcendentals over k. Then $\mathfrak{D} = \{(\theta_r, 0), (\theta_s, 0)\}$, where $k(\theta_r, \theta_s)$ is of degree 30 over $k(f_0, \ldots, f_6)$. It can be checked that the $\xi_i(\mathfrak{D})\,\xi_j(\mathfrak{D})$ $(i \leqslant j)$ are linearly independent over the same field. Hence at most one polynomial of the required kind can specialize to (2) under $\mathfrak{B} \mapsto \mathfrak{D}$. It turns out that we can always find a polynomial B_{ij} which specializes. Further, it is symmetric in $\mathfrak{A}$ and $\mathfrak{B}$. This is reassuring, because we treated the two divisors very differently.

5   Another group property on the jacobian that remains meaningful
on the Kummer is multiplication by an $n \in \mathbb{Z}$. To find $2\mathfrak{A}$ put $\mathfrak{B} = \mathfrak{A}$
in (4.1). Then $\mathfrak{A} - \mathfrak{B} = \mathfrak{O}$, and $\xi_1(\mathfrak{O}) = \xi_2(\mathfrak{O}) = \xi_3(\mathfrak{O}) = 0$. Hence¶ we
have $B_{ij}(\mathfrak{A}, \mathfrak{A}) = 0$ for $i \neq 4$, $j \neq 4$ and we may take

$$\xi_j(2\mathfrak{A}) = B_{j4}(\mathfrak{A}, \mathfrak{A}) \ (1 \leqslant j \leqslant 3), \qquad \xi_4(2\mathfrak{A}) = B_{44}(\mathfrak{A}, \mathfrak{A})/2. \qquad (3.5.1)$$

Then for $n > 2$ one can find the $\xi_j(n\mathfrak{A})$ by induction† using (4.1) with
$\mathfrak{B} = (n-1)\mathfrak{A}$.

6   **A numerical example.** More generally (4.1) enables us to find $\mathfrak{A} \pm \mathfrak{B}$
for given $\mathfrak{A}$, $\mathfrak{B}$, though it cannot tell which is which. Consider a numerical
example. The polynomial

$$F(X) = X^6 - X^5 + 3X^4 - 5X^3 - 10X + 16 \qquad (3.6.1)$$

is chosen so as to be a square for several integral values of X. Consider
$\mathfrak{A} = \{(-1, 6), (0, -4)\}$ and $\mathfrak{B} = \{(1, 2), (2, -6)\}$. Corresponding values of
the Kummer coordinates ξ_j are $(1, -1, 0, 90)$ and $(1, 3, 2, 24)$. Substituting
these in the forms B_{ij} of Theorem 4.1, we get the symmetric table

0	19008	29376	171072
19008	49104	32400	460944
29376	32400	−17136	320976
171072	460944	320976	4319568

$$(3.6.2)$$

Here the (i, j) entry is of the form $a_i b_j + a_j b_i$, where (a_1, a_2, a_3, a_4) and
(b_1, b_2, b_3, b_4) are the coordinates of $\mathfrak{A} \pm \mathfrak{B}$. Removing common factors we
get $\mathbf{a} = (0, 11, 17, 99)$ and $\mathbf{b} = (24, 31, -7, 303)$. The X-coordinates of the
support of the divisors belonging to $\mathbf{a}$ are ∞ and $17/11$. As a check one
verifies that $F(17/11) = (2666/1331)^2$. The divisors belonging to $\mathbf{b}$ are
more typical. The X-coordinates are roots of $24X^2 - 31X - 7$ and so are
$(31 \pm \sqrt{1633})/48$. Here $F((31 + \sqrt{1633})/48) = (5(7015 - 407\sqrt{1633})/2768)^2$.

¶ In checking this one must remember that the $\xi_j(\mathfrak{A})$ satisfy the defining
equation of the Kummer.

† cf. Cassels (1991), Chapter 26, Exercise 2.

7   The tropes. In the classical theory over an algebraically closed ground field, a Kummer surface is projectively self-dual. The dual of the nodes are 16 planes, the *tropes*. Every node is incident with precisely six tropes and *vice versa*: and the nodes and tropes form a structure with interesting properties which are much discussed in Hudson (1905). As we shall soon see, the Kummer is not necessarily self-dual over a general ground field. We can, and shall, however compute the equations of the tropes over $\bar{k}$.

Six tropes are easily found from inspection. For given i the nodes N_0 and N_{ij} all lie on the plane

$$\theta_i{}^2 X_1 - \theta_i X_2 + X_3 = 0. \tag{3.7.1}$$

We denote this trope by T_i: in terms of the dual coordinates it is given by

$$T_i : \quad (\theta_i{}^2, -\theta_i, 1, 0). \tag{3.7.2}$$

The remaining ten tropes correspond to the unordered partitions of the set $\{1, \ldots, 6\}$ into two sets of three, say¶ $\{i, j, k\}$ and $\{l, m, n\}$. Put

$$G(X) = (X - \theta_i)(X - \theta_j)(X - \theta_k) = \sum_{r=0}^{3} g_r X^r,$$
$$H(X) = (X - \theta_l)(X - \theta_m)(X - \theta_n) = \sum_{r=0}^{3} h_r X^r, \tag{3.7.3}$$

so

$$F(X) = f_6 G(X) H(X). \tag{3.7.4}$$

Then $f_6\{G(x)H(u) + G(u)H(x)\} - F_0(x, u)$ is clearly divisible by $(x - u)^2$; and an easy calculation shows that the quotient is

$$f_6\{(g_2 h_0 + g_0 h_2) + (g_0 + h_0)(x + u) + (g_1 + h_1)xu\};$$

that is

$$\frac{f_6\{G(x)H(u) + G(u)H(x)\} - 2yv}{(x - u)^2} \tag{3.7.5}$$
$$= \beta_0 + f_6(g_2 h_0 + g_0 h_2)\sigma_0 + f_6(g_0 + h_0)\sigma_1 + f_6(g_1 + h_1)\sigma_2.$$

But the left hand side of (5) clearly vanishes at

$$\{(x, y), (u, v)\} = \{(\theta_r, 0), (\theta_s, 0)\} \tag{3.7.6}$$

¶ The occasional use of k as an index should not be confused with its standard meaning as the ground field.

whenever
$$\{r, s\} \subset \{i, j, k\} \quad \text{or} \quad \{l, m, n\}. \tag{3.7.7}$$
Hence the plane $T_{ijk} = T_{lmn}$ with dual coordinates
$$(f_6(g_2 h_0 + g_0 h_2), f_6(g_0 + h_0), f_6(g_1 + h_1), 1) \tag{3.7.8}$$
(where the g_j, h_j are given by (2)) passes through the six N_{rs} given by (7). It is therefore a trope.

8   The jacobian as double cover. The jacobian and the Kummer are both projective varieties. To avoid possible confusion, we must first look more nearly at the way that the jacobian is defined and at the nature of the map from the jacobian to the Kummer.

In constructing the jacobian, we started with an affine variety given by 16 basis elements, one of which was σ_0 with the definition $\sigma_0 = 1$. The jacobian is the projective variety given by the ratios of 16 coordinates, which were identified with the 16 basis elements, and which satisfy all the homogeneous relations satisfied by the base elements. In particular, the jacobian contains the projective point for which all the variables are zero, except for that identified with δ. This is the zero of the group law.

The 16 basis elements z_j for the jacobian are of two kinds. There are six elements, the α_j and the γ_j, which change sign when the divisor changes sign. The remaining ten are unaffected. We call these respectively *odd* and *even*. Denote, very temporarily, by T the projective locus of the ten even basis elements. The map from the jacobian to T is everywhere defined: the squares of the odd elements are equal to quadratic forms in the even elements, so if the even elements all vanish, then so also would the odd ones. The map from the jacobian to T is biregular except at the 2-division point.

The Kummer is the projective locus of the four basis elements $\xi_1 = \sigma_0$, $\xi_2 = \sigma_1$, $\xi_3 = \sigma_2$, $\xi_4 = \beta_0$. With the convention $\sigma_0 = 1$, it is easily checked that the space spanned by the ten even base elements is precisely the space spanned by products of two ξ_j. Hence the Kummer is biregularly equivalent to T.

Since the jacobian and the Kummer are both projective, we may multiply the coefficients in each by different nonzero constants. It is convenient to choose these constants so that
$$\xi_1{}^2 = \sigma_0, \quad \xi_1 \xi_2 = \sigma_1, \quad \xi_1 \xi_3 = \sigma_2, \quad \dots, \quad \xi_4{}^2 = \delta. \tag{3.8.1}$$

This coincides with our earlier convention when $\sigma_0 = 1$. What is more useful, we now have $\delta = 1$ when $\xi_4 = 1$.

Now let ν be a linear combination of the six odd base elements of the jacobian. Then ν^2 is even, and is equal to a quadratic form in the ten even base elements: and so to a quartic form in the ξ_j with the identification (1). The value of ν thus distinguishes between the two sheets of the jacobian over $\mathcal{K}$ except on the divisor of zeros of ν. But the only common zeros of all the ν's are the nodes; so we may always distinguish the sheets.

With a later application in mind, we pursue this a little further. Let ν_1, ν_2 be two distinct choices for ν and let W_1 and W_2 be the corresponding forms in the ξ_j. There is also a quartic form W_{12} corresponding to $\nu_1\nu_2$. Then

$$W_1 W_2 = W_{12}^2 \tag{3.8.2}$$

on $\mathcal{K}$. It follows that the zero locus of W_1 is a divisor taken twice.¶

9   The group law on the jacobian. We first extend the results of Section 4:

Lemma 3.9.1. *With the convention (8.1) there are bilinear forms Φ_{ij} such that*

$$\Big(\xi_i(\mathfrak{A} - \mathfrak{B})\,\xi_j(\mathfrak{A} + \mathfrak{B})\Big) = \Big(\Phi_{ij}\big(\mathbf{z}(\mathfrak{A}), \mathbf{z}(\mathfrak{B})\big)\Big). \tag{3.9.1}$$

Here $\mathbf{z} = (z_0, \ldots, z_{15}) = (\delta, \ldots, \rho)$ of Chapter 2. If the coordinates are normalized so that $\xi_1(\mathfrak{A} \pm \mathfrak{B}) = \sigma_0(\mathfrak{A}) = \sigma_0(\mathfrak{B}) = 1$, then the coefficients of the Φ_{ij} are in $\mathbb{Z}[f_0, \ldots, f_6]$

Proof. We could construct the forms using the local expansions in the neighbourhood of $\mathfrak{O}$. The local group law gives the expansion of the left hand side. Since the dominant terms in the local expansions of the base elements z_k are distinct by (2.3.4), it is then straightforward to find the coefficients in the bilinear forms.

Alternatively, the left hand side of (1) is unchanged on changing the signs of $\mathfrak{A}$ and $\mathfrak{B}$ simultaneously. Hence the pairs of basis elements on the right hand side with nonzero elements must be either even-even or odd-odd. In particular,

$$\Big(\xi_i(\mathfrak{A} - \mathfrak{B})\,\xi_j(\mathfrak{A} + \mathfrak{B}) + \xi_i(\mathfrak{A} + \mathfrak{B})\,\xi_j(\mathfrak{A} - \mathfrak{B})\Big) = \Big(\Phi_{ij} + \Phi_{ji}\Big) \tag{3.9.2}$$

consists only of the even-even terms. But the left hand side of (2) is just the left hand side of (4.1). Hence we know the even-even terms. There only

---

¶ These are the 'octavic curves through sixteen nodes' of Hudson (1905), Section 93. It is shown there that $W_1 W_2 - W_{12}^2 = KL$, where $K = 0$ is the equation of $\mathcal{K}$ and $L = 0$ is the equation of another Kummer surface.

six odd basis elements, so the determination of the remaining coefficients is easier. One can, for example, put $\mathfrak{B} = \pm\mathfrak{A}$ and equate coefficients.

THEOREM 3.9.1. *There are biquadratic forms* Ψ_{IJ} *for* $0 \leqslant I \leqslant 15$, $0 \leqslant J \leqslant 15$ *such that*

$$\left(z_I(\mathfrak{A} - \mathfrak{B}) z_J(\mathfrak{A} + \mathfrak{B}) \right) = \left(\Psi_{IJ}\big(\mathbf{z}(\mathfrak{A}), \mathbf{z}(\mathfrak{B})\big) \right). \qquad (3.9.3)$$

If the coordinates of $\mathfrak{A}$, $\mathfrak{B}$, $\mathfrak{A} \pm \mathfrak{B}$ *are all normalized to* $\sigma_0 = 1$, *then the coefficients in the* Ψ_{IJ} *are in* $\mathbb{Z}[f_0, \ldots, f_6]$. *With the same normalization,*

$$\Psi_{IJ}(\mathbf{o}, \mathbf{o}) = \begin{cases} 1 & \text{if } I = J = 0, \\ 0 & \text{otherwise,} \end{cases} \qquad (3.9.4)$$

where $\mathbf{o} = (1, 0, 0, \ldots, 0)$ *is the coordinates of* $\mathfrak{O}$.

Proof. For the I, J such that z_I and z_J are both even basis elements, this follows immediately by multiplying the identities (1) in pairs and then expressing products such as $\xi_{i_1}(\mathfrak{A} + \mathfrak{B}) \, \xi_{i_2}(\mathfrak{A} + \mathfrak{B})$ in terms of the even z_I: and similarly for $\mathfrak{A} - \mathfrak{B}$. When either z_I or z_J is an odd basis element, much more work is required. The square of the left hand side can then be expressed in terms of the even basis elements z_j, and then one takes the square root of the resulting function of $\mathbf{z}(\mathfrak{A})$ and $\mathbf{z}(\mathfrak{B})$.

The theorem gives not one, but 16 distinct addition theorems, but none of them is everywhere defined. By the theorem, for fixed I the vector

$$\left(\Psi_{I\,0}\big(\mathbf{z}(\mathfrak{A}), \mathbf{z}(\mathfrak{B})\big), \Psi_{I\,1}\big(\mathbf{z}(\mathfrak{A}), \mathbf{z}(\mathfrak{B})\big), \ldots, \Psi_{I\,15}\big(\mathbf{z}(\mathfrak{A}), \mathbf{z}(\mathfrak{B})\big) \right) \qquad (3.9.5)$$

is projectively equivalent to $\mathbf{z}(\mathfrak{A} + \mathfrak{B})$ provided that not all the entries are 0; that is, provided that $z_I(\mathfrak{A} - \mathfrak{B}) \neq 0$. In particular, for $\mathfrak{A} = \mathfrak{B}$ one must take $I = 0$:

Corollary. $2\mathfrak{A}$ *has coordinates*

$$\left(z_J(2\mathfrak{A}) \right) = \left(\Psi_{0J}\big(\mathbf{z}(\mathfrak{A}), \mathbf{z}(\mathfrak{A})\big) \right). \qquad (3.9.6)$$

Again, since the local parameters are z_1/z_0 and z_2/z_0, the formal group law at $\mathfrak{O}$ is immediately derivable from the theorem: this process will be discussed in more detail in Chapter 7.

10   Recovery of the curve. A classical theorem over an algebraically closed field asserts that every abelian variety of dimension 2 is isogenous to the jacobian of a curve. Indeed the curve shows up in the geometry of the Kummer in many ways. For example, the Kummer meets a trope in a conic (taken twice) and the six nodes incident with the trope are projective with the roots of the corresponding $F(X)$. Again, it can be shown that a tangent plane meets the Kummer in a curve birationally equivalent to $\mathcal{C}$. [Cf. Chapter 4, Section 4.]

For non-algebraically-closed fields we must pay attention to fields of definition. Consider first the Kummer of a curve defined over the ground field ('rational') as we have constructed it in this chapter. At least one of the nodes is rational, namely $N_0 = (0, 0, 0, 1)$. As we have seen in display (1.8), the equation $K = 0$ of $\mathcal{K}$ has the shape

$$K = K_2\xi_4{}^2 + K_3\xi_4 + K_4, \qquad (3.10.1)$$

where the K_j are rational forms in ξ_1, ξ_2, ξ_3 of degree j: in particular,

$$K_2 = \xi_2{}^2 - 4\xi_1\xi_3. \qquad (3.10.2)$$

A line through N_0 is given by the ratios of ξ_1, ξ_2, ξ_3 (not all 0): it meets $\mathcal{K}$ in at most one further point¶ precisely when $K_2 = 0$. Hence the surface $K_2 = 0$, which is a cone with vertex N_0, is intrinsically defined by the geometry: it is called the tangent cone at N_0. The lines through N_0 are the rays of the cone. Note that the quadratic form K_2 represents 0 nontrivially, and so there are rational rays.

We have also seen that over $\bar{k}$ there are six tropes through N_0, namely $\theta_j{}^2\xi_1 - \theta_j\xi_2 + \xi_3 = 0$. Interpreting ξ_1, ξ_2, ξ_3 as coordinates in $\mathbb{P}^2$, these are tangents to the conic $K_2 = 0$ at the points $(1, 2\theta_j, \theta_j{}^2)$.

We are now in a position to state the main result of this section. We shall mean by an 'abstract Kummer' an algebraic surface of degree 4 in $\mathbb{P}^3$ with 16 nodes and 16 tropes. 'Rational' means 'defined over the ground field'.

THEOREM 3.10.1. *Let $\mathcal{K}$ be a rational abstract Kummer. Suppose that it has a rational node N_0 and that the tangent cone at N_0 has a rational ray. Then $\mathcal{K}$ is the Kummer of a rational curve $\mathcal{C} : Y^2 = F(X)$ of genus 2.*

We have just proved that the conditions are necessary. To prove sufficiency we may take the coordinates of N_0 as $(0, 0, 0, 1)$. The equation of $\mathcal{K}$ is now of the form

$$K(\mathbf{Y}, Z) = K_2(\mathbf{Y})Z^2 + K_3(\mathbf{Y})Z + K_4(\mathbf{Y}) = 0, \qquad (3.10.3)$$

¶ With the usual theology about points 'counted twice'.

where $\mathbf{Y} = (Y_1, Y_2, Y_3)$ and K_j is a rational form of degree j. The form K_2 is nonsingular: by hypothesis it has a rational zero. Hence without loss of generality (multiplying K by a rational constant if necessary)

$$K_2 = Y_2^2 - 4Y_1 Y_3. \tag{3.10.4}$$

There are six tropes through N_0, and they are tangent to the cone $K_2 = 0$, say at $(Y_1, Y_2, Y_3) = (1, 2\theta_j, \theta_j{}^2)$ for $(j = 1, \ldots, 6)$. [We can avoid $\theta = \infty$ by suitable choice of the coordinates in (4).] The six tropes are thus

$$T_j : \qquad \theta_j{}^2 Y_1 - \theta_j Y_2 + Y_3 = 0. \tag{3.10.5}$$

Any line in a trope meets $\mathcal{K}$ only in double points. Hence the discriminant of (1) with respect to Z vanishes on the six tropes (3), and so

$$K_3^2 - 4K_2 K_4 = c \prod_j (\theta_j{}^2 Y_2 - \theta_j Y_1 + Y_3). \tag{3.10.6}$$

The left hand side is rational, so c and

$$F(X) = \prod (X - \theta_j) \tag{3.10.7}$$

are rational.

It can be shown¶ that the configuration of 16 lines and 16 planes with the incidence properties of the nodes and tropes of an abstract Kummer surface is determined up to a transformation $Z \mapsto c_1 Y_1 + c_2 Y_2 + c_3 Y_3 + dZ$ by six planes tangent to a quadric cone. Hence the abstract Kummer $\mathcal{K}$ must be (projectively equivalent to) the Kummer of $Y^2 = F(X)$, as required.

¶ Alternatively, Hudson (1905), Section 10 (pp. 17–19) derives the equation of the Kummer from the given data, but without regard to the field of definition.

Chapter 4

The dual of the Kummer

0   Introduction. A reader primarily interested in the computational aspects may omit this chapter and the next at a first reading.

Classically, over an algebraically closed field, the Kummer surface is isomorphic to its projective space dual. This is no longer the case in our situation. In this chapter we construct the dual. We do so not directly, but by solving an apparently quite different problem.

The jacobian of a curve of genus g was first constructed not from Pic^g but from Pic^n for some $n > g$. Here we first obtain a variety which parametrizes Pic^3 modulo the involution induced by the involution $\pm Y$ of the curve $\mathcal{C}$. It turns out to be a quartic surface $\mathcal{K}^*$ in $\mathbb{P}^3$. A direct, if somewhat mysterious, computation¶ shows that $\mathcal{K}^*$ is indeed the required dual. The duality is described in terms of the canonical map from $\mathrm{Pic}^2 \oplus \mathcal{C}$ to Pic^3 which maps the (class of the) divisor $\{\mathfrak{a}, \mathfrak{b}\}$ and the point $\mathfrak{c}$ into the class of the divisor $\{\mathfrak{a}, \mathfrak{b}, \mathfrak{c}\}$.

The chapter ends with a number of miscellaneous applications.

1   Description of Pic^3. We use a construction of Jacobi (1846), see also [Mumford (1983), II, p. 317]. A divisor class defined over the ground field (k, as usual) does not necessarily contain a divisor defined over k.† Hence we work over the algebraic closure, though the final formulae will be in k.

We shall say that an effective divisor $\mathfrak{A}$ on $\mathcal{C}$ of degree 3 is in general position if it is given by

$$U(X) = 0, \qquad Y = W(X), \tag{4.1.1}$$

where U and W have degree at most 3. If U has degree < 3, then some of the points of the support are at infinity. It is easy to see that $\mathfrak{A}$ is not in general position precisely when either (i) there are distinct points $\mathfrak{a}$, $\mathfrak{b}$ of

---

¶ A proof on different lines is in Chapter 5.

† As we shall see. For the background, cf. Appendix to this chapter.

its support such that $\{\mathfrak{a}, \mathfrak{b}\}$ is in the canonical class $\mathfrak{O}$ or (ii) a Weierstrass point $(\theta, 0)$ occurs with multiplicity > 1. For given $\mathfrak{A}$, the polynomial U is determined up to a scalar nonzero factor and W is determined modulo U.

With the same notation, there is a $V(X)$ of degree at most 3 such that

$$F(X) = W(X)^2 - U(X)V(X). \tag{4.1.2}$$

The effective divisor $\mathfrak{B}$ of degree 3 given by

$$V(X) = 0, \qquad Y = -W(X) \tag{4.1.3}$$

is also in general position. Further, $\mathfrak{A} - \mathfrak{B}$ is the divisor of

$$\frac{W(X) - Y}{V(X)} = \frac{U(X)}{W(X) + Y}; \tag{4.1.4}$$

so $\mathfrak{A}$ and $\mathfrak{B}$ are linearly equivalent.

By Riemann-Roch, any divisor $\mathfrak{A}_1$ in the divisor class other than $\mathfrak{B}$ is the divisor of zeros of

$$\frac{W(X) - Y}{V(X)} + c = \frac{(W + cV) - Y}{V} \tag{4.1.5}$$

for some scalar c. Here

$$F = (W + cV)^2 - V(U + 2cW + c^2 V). \tag{4.1.6}$$

Similarly, we may replace $\mathfrak{B}$ by a divisor $\mathfrak{B}_1$ in the class other than $\mathfrak{A}_1$.

In this way we arrive at another representation

$$F = W_1^2 - U_1 V_1, \tag{4.1.7}$$

where U_1, V_1, W_1 come from U, V, W by a linear transformation with constant coefficients and determinant $+1$. Further, every¶ proper† automorph of $W^2 - UV$ is obtainable in this way. Hence the linear equivalence classes of effective divisors of degree 3 in general position are in 1–1 correspondence with the proper equivalence classes of representations of F by $W^2 - UV$, where U, V, W have degree at most 3.

Note also that the involution $Y \mapsto -Y$ corresponds to $U \mapsto U$, $V \mapsto V$, $W \mapsto -W$. Hence an improper automorph of $W^2 - UV$ takes the corresponding element of Pic^3 into its conjugate.

¶ Recall that we are working over the algebraic closure.

† i.e. of determinant $+1$.

2 We now introduce new variables X_0, X_1, X_2, X_3 and linear forms U_l, V_l, W_l in them which go into U, V, W under

$$X_j \longmapsto X^j \qquad (j = 0, 1, 2, 3). \tag{4.2.1}$$

The quadratic form

$$S = S(X_0, X_1, X_2, X_3) = W_l^2 - U_l V_l \tag{4.2.2}$$

is unchanged under automorphs of $W^2 - UV$, so S depends only on the element of Pic^3 to which the divisor (1.1) belongs, and is fixed under the involution of Pic^3 induced by $Y \mapsto -Y$. In particular, if the element of Pic^3 is rational, then so is S.

The quadratic form

$$S_4 = f_0 X_0^2 + f_1 X_0 X_1 + f_2 X_1^2 + f_3 X_1 X_2 + f_4 X_2^2 + f_5 X_2 X_3 + f_6 X_3^2 \tag{4.2.3}$$

goes into $F(X)$ under (1). A basis for the kernel of (1) acting on the space of quadratic forms is

$$\begin{aligned} S_1 &= X_0 X_2 - X_1^2, \\ S_2 &= X_0 X_3 - X_1 X_2, \\ S_3 &= X_1 X_3 - X_2^2. \end{aligned} \tag{4.2.4}$$

Hence the equation $F = W^2 - UV$ is equivalent to

$$W_l^2 - U_l V_l = S = \sum_{1}^{4} \eta_j S_j, \tag{4.2.5}$$

where

$$\begin{aligned} \eta_1 &= 2w_0 w_2 - u_0 v_2 - u_2 v_0, \\ \eta_2 &= - u_0 v_3 - u_3 v_0, \\ \eta_3 &= - u_1 v_3 - u_3 v_1, \\ \eta_4 &= 1. \end{aligned} \tag{4.2.6}$$

Here u_j is the coefficient of X^j in U, etc. In particular, the quadratic form (5) is singular, and so $K^*(\eta_1, \eta_2, \eta_3, \eta_4) = 0$, where

$$K^* = \begin{vmatrix} 2f_0\eta_4 & f_1\eta_4 & \eta_1 & \eta_2 \\ f_1\eta_4 & 2f_2\eta_4 - 2\eta_1 & f_3\eta_4 - \eta_2 & \eta_3 \\ \eta_1 & f_3\eta_4 - \eta_2 & 2f_4\eta_4 - 2\eta_3 & f_5\eta_4 \\ \eta_2 & \eta_3 & f_5\eta_4 & 2f_6\eta_4 \end{vmatrix} \tag{4.2.7}$$

is a form of degree 4 in the η_j. We denote the variety $K^* = 0$ by $\mathcal{K}^*$.

The argument so far is reversible. Given a point of $\mathcal{K}^*$ with $\eta_4 \neq 0$, we may take $\eta_4 = 1$ by homogeneity. The quadratic form $\sum \eta_j S_j$ is singular, and so¶ is representable as $W_l^2 - U_l V_l$. The transformation $X_j \mapsto X^j$ now gives an effective divisor of degree 3 on C. Further, the class of the divisor up to linear equivalence and the $\pm Y$ involution depends only on the η_j. Hence the corresponding element of Pic^3 is either defined over k or defined over a quadratic extension, its conjugate over k being its conjugate under the $\pm Y$ involution.

It remains to find an interpretation of the points of $\mathcal{K}^* = 0$ with $\eta_4 = 0$, so $K^* = (\eta_2{}^2 - \eta_1 \eta_3)^2$. For this, we must consider the effective divisors of degree 3 which are not in general position as defined above, that is, are of the form $\mathfrak{A} = \mathfrak{b} \oplus \mathfrak{C}$, where $\mathfrak{b}$ is a point on C and $\mathfrak{C}$ is in the canonical class $\mathfrak{D}$. Any divisor linearly equivalent to $\mathfrak{A}$ is of the same form with the *same* $\mathfrak{b}$. We have an injection of Pic^1 into Pic^3. If $\mathfrak{b} = (x, y)$, a specialization argument with a divisor 'near' $\mathfrak{C}$ shows that the appropriate class of singular quadratic forms is that of

$$(X_2 - xX_1)^2 - (X_3 - xX_2)(X_1 - xX_0) = -x^2 S_1 + x S_2 - S_3. \qquad (4.2.8)$$

The relevant point on $\mathcal{K}^*$ is thus $(\eta_1, \eta_2, \eta_3, \eta_4) = (x^2, -x, 1, 0)$, with the obvious interpretation if $\mathfrak{b}$ is at infinity.

To sum up:

THEOREM 4.2.1. *The points on the projective variety $\mathcal{K}^*$ classify the effective divisors of degree 3 on C up to linear equivalence and the $\pm Y$ involution.*

3 $\mathcal{K}^*$ is the projective dual of $\mathcal{K}$. Recall that the points of $\mathcal{K}$ classify the effective divisors of degree 2 on C up to linear equivalence and the $\pm Y$ involution. We use the coordinates ξ_j introduced in Chapter 3.

THEOREM 4.3.1. *Let $\xi = (\xi_1, \xi_2, \xi_3, \xi_4)$ and $\eta = (\eta_1, \eta_2, \eta_3, \eta_4)$ be points of $\mathcal{K}$ and $\mathcal{K}^*$ respectively. A necessary and sufficient condition that*

$$\xi_1 \eta_1 + \xi_2 \eta_2 + \xi_3 \eta_3 + \xi_4 \eta_4 = 0 \qquad (4.3.1)$$

is that there exist effective divisors $\mathfrak{A}$, $\mathfrak{B}$ of degrees 2, 3, belonging to ξ, η, and a point $\mathfrak{x}$ of C such that

$$\mathfrak{B} = \mathfrak{A} \oplus \mathfrak{x}. \qquad (4.3.2)$$

¶ Recall again we are working over the algebraic closure.

Note 1. Cognoscenti will recognize Θ-divisors, so the duality is also one in the sense of abelian varieties. We explore this later.

Note 2. The proof given here is a direct algebraic verification. We shall give a more conceptual proof in Chapter 5.

Proof. If $\xi = (0,0,0,1)$, then $\mathfrak{A} \in \mathfrak{O}$ and we proved the assertion of the theorem at the end of the last section. We may thus assume that $\mathfrak{A} \notin \mathfrak{O}$. We shall show, first, that (2) implies (1). We shall assume, further, that neither $\mathfrak{x}$ nor $\bar{\mathfrak{x}}$ is in the support of $\mathfrak{A}$: the excluded cases follow by an obvious specialization argument. By a suitable choice of coordinates for $\mathcal{C}$ we may thus take

$$F(X) = W(X)^2 - X(X - x)(X - u)V(X) \qquad (4.3.3)$$

for nonzero x, u and cubics $W(X)$, $V(X)$. Here

$$\begin{aligned}
\mathfrak{x} &= \big(0, W(0)\big), \\
\mathfrak{A} &= \big((x, W(x)), (u, W(u))\big) \\
\mathfrak{B} &= \big((x, W(x)), (u, W(u)), (0, W(0))\big).
\end{aligned} \qquad (4.3.4)$$

For $\eta(\mathfrak{B})$ we have only to put $U(X) = X(X - x)(X - u)$ in (2.6):

$$\begin{aligned}
\eta_1 &= 2w_0 w_2 + (x + u)v_0, \\
\eta_2 &= 2w_0 w_3 - v_0, \\
\eta_3 &= 2w_1 w_3 - v_1 - xuv_3, \\
\eta_4 &= 1
\end{aligned} \qquad (4.3.5)$$

where w_j is the coefficient of $X^j W(X)$. For $\xi(\mathfrak{A})$ we have

$$\begin{aligned}
\xi_1 &= \sigma_0(\mathfrak{A}) = 1, \\
\xi_2 &= \sigma_1(\mathfrak{A}) = x + u, \\
\xi_3 &= \sigma_2(\mathfrak{A}) = xu, \\
\xi_4 &= \beta_0(\mathfrak{A}) = \frac{F_0(x, u) - 2W(x)W(u)}{(x - u)^2} \\
&= v_3 x^2 u^2 + v_1 xu - 2w_0 w_3 - 2(x + u)w_0 w_3 - 2xuw_1 w_3,
\end{aligned} \qquad (4.3.6)$$

after some calculation.¶

It is now straightforward that (1) holds. For fixed $\mathfrak{A}$ we thus have an algebraic map from $\mathcal{C}$ into a plane section of $\mathcal{K}^*$. Since the latter is irreducible and has the same dimension as $\mathcal{C}$, the map is onto. This concludes the proof of the theorem.

¶ Special case of divisor $G(X) = 0$, $Y = W(X)$, where:
$F(X) = W(X)^2 - G(X)H(X)$ and $G(X), W(X), H(X)$ have degrees 2,3,4.

We can identify the tropes of $\mathcal{K}$ (= nodes of $\mathcal{K}^*$) and confirm their coordinates, which we have already obtained in Chapter 3, Section 7. Ten tropes are given by the quadratic forms of rank 2 linearly dependent on the S_j: for in the situation discussed in Sections 1, 2 they correspond to factorizations $F = -UV$, where $U, V \in \bar{k}[X]$ are cubics. On putting $U(X) = f_6 G(X)$, $V(X) = -H(X)$ and $W(X) = 0$ in (2.6) we recover (3.7.8). The remaining six tropes T_j of $\mathcal{K}$ are given by (2.8) with $x = \theta_j$ in accordance with (3.7.2).

It appears to have been an entirely uncovenanted mercy that our quite separate constructions of $\mathcal{K}$ and $\mathcal{K}^*$ should lead naturally to dual bases.

4   Tangency. With a little more trouble, we can establish the correspondence given by tangency between the points of $\mathcal{K}$ and $\mathcal{K}^*$.

Let the notation be as in Theorem 3.1. For fixed $\mathfrak{A}$ (and so fixed ξ), (3.1) is a plane in the dual space $\mathbb{P}^{3*}$. By the theorem, the points η of its intersection with $\mathcal{K}^*$ correspond to the classes of the divisors $\mathfrak{A} \oplus \mathfrak{x}$, where $\mathfrak{x} \in \mathcal{C}$. Two distinct points $\mathfrak{x}_1$, $\mathfrak{x}_2$ of $\mathcal{C}$ give rise to the same point of the section precisely when $\mathfrak{A} \oplus \mathfrak{x}_1$ is linearly equivalent either to $\mathfrak{A} \oplus \mathfrak{x}_2$ or to $\overline{\mathfrak{A} \oplus \mathfrak{x}_2}$. The first is impossible, the second gives

$$2\mathfrak{A} = \{\bar{\mathfrak{x}}_1, \bar{\mathfrak{x}}_2\} \tag{4.4.1}$$

in the sense of the jacobian group law. If $2\mathfrak{A} = \mathfrak{O}$, this is satisfied by $\mathfrak{x}_2 = \bar{\mathfrak{x}}_1$: we have a trope of $\mathcal{K}^*$ and the section is a conic (taken twice). Otherwise, $\{\mathfrak{x}_1, \mathfrak{x}_2\}$ is uniquely determined by $\mathfrak{A}$. The section is a plane quartic birationally equivalent to $\mathcal{C}$ with a double point corresponding to the pair of points $\mathfrak{x}_1$, $\mathfrak{x}_2$. The double point is, of course, the point of tangency of the plane (3.1) to $\mathcal{K}^*$.

The tangents to $\mathcal{K}$ can be treated similarly. Let η and a divisor $\mathfrak{B}$ of degree 3 belonging to it be fixed. Then the ξ on (3.1) belong to the divisors $\mathfrak{C}$ of degree 2 such that $\mathfrak{B} \oplus \eta$ is linearly equivalent to $\mathfrak{C} \oplus \mathfrak{O}$ as the point η runs over $\mathcal{C}$. Two distinct points η_1 and η_2 give the same $\mathfrak{C}$ precisely when $\mathfrak{B} \oplus \mathfrak{B}$ is linearly equivalent to $\eta_1 \oplus \eta_2 \oplus \mathfrak{O}$.

Combining the two previous paragraphs, we have a birational correspondence between $\mathcal{K}$ and $\mathcal{K}^*$. If $\mathfrak{B} = \mathfrak{A} \oplus \mathfrak{x}_1$ and $\eta_1 = \bar{\mathfrak{x}}_1$ then $\mathfrak{C} = \mathfrak{A}$ and $\eta_2 = \mathfrak{x}_2$. [Note, no bar this time.]

Alternatively,¶ one can find an explicit description of the tangent to $\mathcal{K}$ as follows. Let $\mathfrak{A} \not\equiv \mathfrak{O}$ be an effective divisor of degree 2. Without loss of generality its support is in the finite part of the plane. Let $M(X)$ be the cubic uniquely determined by the condition that $Y = M(X)$ meets $\mathcal{C}$ in $\mathfrak{A}$

¶ The rest of this section can be omitted at first reading.

with multiplicity 2. Then

$$F(X) = M(X)^2 + G(X)^2 H(X), \qquad (4.4.2)$$

where $G(X), H(X)$ are quadratics and $\mathfrak{A}$ is $G(X) = 0$, $Y = M(X)$. Let $\mathfrak{X} = \{(x, y), (u, v)\}$ be a generic divisor and put

$$\widetilde{F_0(x, u)} = 2M(x)M(u) + G(x)G(u)H_0(x, u), \qquad (4.4.3)$$

where $H_0(x, u) = 2h_0 + h_1(x + u) + 2h_2 xu$, with our usual convention for coefficients of polynomials. It is easily seen that

$$\widetilde{F_0(x, u)} - 2vy \qquad (4.4.4)$$

vanishes when $\mathfrak{X}$ is specialized to $\mathfrak{A}$. Further,

$$\frac{\widetilde{F_0(x, u)} - 2yv}{(x - u)^2} - \beta_0 = \frac{\widetilde{F_0(x, u)} - F_0(x, u)}{(x - u)^2} \qquad (4.4.5)$$
$$= H_1 + H_2(x + u) + H_3 xu,$$

where
$$\begin{aligned} H_1 &= 2h_0 g_0 g_2 + h_1 g_0 g_1 + 2m_0 m_2, \\ H_2 &= h_1 g_0 g_2 + 2m_0 m_3, \qquad\qquad (4.4.6) \\ H_3 &= 2h_1 g_1 g_2 + 2h_2 g_0 g_2 + 2m_1 m_3. \end{aligned}$$

Hence the plane
$$H_1 \xi_1 + H_2 \xi_2 + H_3 \xi_3 + H_4 \xi_4 = 0 \qquad (4.4.7)$$

with $H_4 = 1$ passes through the point of $\mathcal{K}$ belonging to $\mathfrak{A}$. A straightforward calculation (preferably with an algebra package) shows that it is the tangent to $\mathcal{K}$ there.

5 Explicit dualities.

Classically, over an algebraically closed field, the Kummer is projective with its dual. This is no longer the case in our setup. A projective duality takes nodes into tropes and *vice versa*. There is always one node, namely that belonging to $\mathfrak{O}$, defined over the ground field but in general none of the tropes is.

THEOREM 4.5.1. *A necessary and sufficient condition that $\mathcal{K}$ and $\mathcal{K}^*$ be projective over the ground field is that at least one trope be defined over that field.*

As already noted, the condition is necessary. To prove sufficiency, we must consider, first, projectivities over the algebraic closure. For a general point $\mathfrak{r}$ of C the map from $\mathcal{K}$ to $\mathcal{K}^*$ given by (3.2) is only 2–2, since a point ξ of $\mathcal{K}$ determines a divisor class $\mathfrak{A}$ only up to the $\pm Y$ involution, and similarly for $\mathcal{K}^*$. When $\mathfrak{r}$ is a Weierstrass point, however, the map becomes 1–1, and is linear, as the following direct calculation shows. Take $\mathfrak{r} = (\theta_6, 0)$, where without loss of generality $\theta_6 = 0$. Let $(6, i, j, l, m, n)$ be a permutation of $(1, 2, 3, 4, 5, 6)$. Then by (3.1.14) the coordinates of the nodes N_{ij} $(i \neq 6,\ j \neq 6)$ are

$$N_{ij}:\quad (\xi_1, \xi_2, \xi_3, \xi_4) =$$
$$\Big(1,\ \theta_i + \theta_j, \theta_i\theta_j,\ -f_6[(\theta_i\theta_j)^2 + (\theta_i\theta_j)(\theta_l\theta_m + \theta_m\theta_n + \theta_n\theta_l)]\Big). \tag{4.5.1}$$

By (3.7.8) the coordinates of the tropes T_{ij6} in dual coordinates are

$$T_{ij6}:\quad (\eta_1, \eta_2, \eta_3, \eta_4) = \Big(f_6[(\theta_i + \theta_j)(\theta_l\theta_m\theta_n)],\ -f_6(\theta_l\theta_m\theta_n),$$
$$f_6[(\theta_i\theta_j) + (\theta_l\theta_m + \theta_m\theta_n + \theta_n\theta_l)], 1\Big). \tag{4.5.2}$$

It is readily checked that they are taken into one another by the transformation

$$\eta_1 : \eta_2 : \eta_3 : \eta_4 = f_1\xi_2 : -f_1\xi_1 : \xi_4 : -\xi_3. \tag{4.5.3}$$

[Note that $f_1 = -f_6 \prod_{i \neq 6} \theta_i$ when $\theta_6 = 0$.] Further, (3) takes N_0 into T_6 and N_{i6} into T_i.

That (3) takes the Kummer into its dual follows from the fact that the Kummer is uniquely defined by its nodes [Hudson (1905), p. 81]. That the matrix of the transformation (3) is skew-symmetric is bound up with the existence of a line-complex associated with a Kummer surface. The line-complex has a major rôle in the classical theory: we will give a brief account in Chapter 17, but otherwise it plays no part in our story.

We have proved

Lemma 4.5.1. *For each Weierstrass point $(\theta_j, 0)$ there is a projective map $L(j)$ from $\mathcal{K}$ to $\mathcal{K}^*$ defined over¶ $k(\theta_j)$ which takes the node N_0 to the trope T_j.*

Corollary. *For $i \neq j$ the map $L(i)^{-1}L(j)$ from $\mathcal{K}$ to itself is the linear map induced by addition of $\{(\theta_i, 0), (\theta_j, 0)\}$.*

The corollary is immediate. It follows that $L(i)^{-1}L(j) = L(j)^{-1}L(i)$. When the projective maps are lifted to maps of the underlying affine spaces it may be checked that they anticommute. This gives another approach to the anticommutation properties of the $W[ij]$ of Chapter 3, Section 3.

¶ k is the ground field.

We can now complete the proof of the theorem. Let l, m, n be distinct indices. Then $L(l)L(m)^{-1}L(n)$ is a projective map which takes $\mathcal{K}$ into $\mathcal{K}^*$ and takes the node N_0 into the trope T_{lmn}. It is uniquely defined by these properties, and so is defined over the ground field when T_{lmn} is.

6   We conclude with a comment on the classical theory, which deals with a curve $Y^2 = F(X)$ with F of degree 5. There is then a Weierstrass point at infinity, and the classical identification between $\mathcal{K}$ and $\mathcal{K}^*$ is the one corresponding to it. In our setup it is not canonical because it depends on the choice of Weierstrass point.

Appendix

Rational divisor classes with no rational divisor

For an easy example in genus 2, let $F(X) = -G(X)^2 - H(X)^2$, where $G(X), H(X) \in \mathbb{Q}[X]$ are of degree 3, 2 respectively and have no common zero. There are no real points on $\mathcal{C} : Y^2 = F(X)$, and complex points occur in conjugate pairs. Hence any rational divisor has even degree. On the other hand, the divisor $G(X) + \sqrt{-1}H(X) = 0, Y = 0$ of degree 3 is linearly equivalent to its conjugate: and so there is a rational divisor class of degree 3.

A similar but simpler argument shows, on the curve $\mathcal{B} : X^2 + Y^2 + 1 = 0$ of genus 0, that Pic^1 is a rational divisor class without rational divisor. For an example in genus 1, see Cassels (1963).

In the other direction, if the curve $\mathcal{B}$ of arbitrary genus is defined over the ground field k and has a nonsingular¶ rational point P, then every rational divisor class contains a rational divisor. For it certainly contains a divisor $\mathfrak{d}$ defined over $\bar{k}$, and we may suppose that P is not in its support. Then $\sigma\mathfrak{d} - \mathfrak{d}$ is principal for any $\sigma \in \mathrm{Gal}$, say $[\phi_\sigma] = \sigma\mathfrak{d} - \mathfrak{d}$ for some function ϕ_σ. Without loss of generality, $\phi_\sigma(P) = 1$. The coboundary of $\{\phi_\sigma\}$ takes values in $\bar{k}$, and so is trivial. By Hilbert 90 applied to the function field, $\phi_\sigma = \sigma\psi/\psi$ for some function ψ. Then $[\psi] = \mathfrak{d} - \mathfrak{e}$ for some $\mathfrak{e}$ invariant under Galois, i.e. defined over k. If k is a number field, then it is enough that $\mathcal{B}$ has a point everywhere locally, by the local-global principle for the Brauer group.

For a fuller discussion see Coray & Manoil (1995).

¶ This proviso may be omitted if 'point' is taken to mean 'place'.

Chapter 5

Weddle's surface

0   Introduction. This chapter may be omitted at a first reading.

In Chapter 3 we introduced the Kummer surface as the locus of some even elements of the basis for the jacobian constructed in Chapter 2. Here we consider the locus of odd elements

$$(\alpha_0, \alpha_1, \alpha_2, \alpha_3). \tag{5.0.1}$$

This turns out to be yet another quartic surface $\mathcal{W}$, but this time with only six nodes. In the classical situation, it is called Weddle's surface,¶ and was much studied in the 19th century.

We shall not have much occasion to use Weddle's surface, but the discussion is an elegant piece of old-fashioned geometry which follows on naturally from Chapter 4. As a bonus, it gives another proof of the projective duality of $\mathcal{K}$ and $\mathcal{K}^*$.

The old literature [Jessop (1916), Chapter 9, especially Section 107; Hudson (1905), Chapter 15; Baker (1907), pp. 37, 38] gives a birational correspondence between Weddle's surface and the Kummer. We have to respect the ground field, and it turns out that the natural correspondence is with the dual $\mathcal{K}^*$. It is readily verified that all the divisors $\{(\theta, 0), \mathfrak{x}\}$, where θ is a root of $F(X)$, map under (1) into the same point $(1, \theta, \theta^2, \theta^3)$. This gives the six nodes of $\mathcal{W}$: they correspond to the six tropes T_j of $\mathcal{K}$. The remaining ten nodes of $\mathcal{K}^*$ blow up into lines on $\mathcal{W}$. As this chapter

---

¶ It appears first on p. 69 of the *Cambridge and Dublin Mathematical Journal* **5** (1850) in a footnote which must be quoted:

'I cannot forbear expressing a doubt, though with some diffidence, as to the correctness of one of the theorems which M. Chasles has given in Note XXXIII. ... Now the analysis to which I have subjected the problem leads me to conclude that the locus is a *surface* of the fourth degree ....'

When he wrote, Thomas Weddle was Mathematical Master at the National Society's Training College, Battersea. He became Mathematics Professor at the Royal Military College at Sandhurst in 1851, but died in 1853. Born 1817. He did not make the *Dictionary of Scientific Biography* but is in Poggendorf, *Handwörterbuch der exakten Wissenschaften*.

is already a digression, we shall not discuss the geometry of $\mathcal{W}$ further. [Cf. table on p. 167 of Hudson (1905).]

1   Symmetroid and Jacobian. In this section we introduce two quartic surfaces associated with four sufficiently general quadratic forms $\mathbf{S} = (S_1, S_2, S_3, S_4)$ in $\mathbf{X} = (X_1, X_2, X_3, X_4)$. The Jacobian¶ $\partial(\mathbf{S})/\partial(\mathbf{X})$ is the determinant of the matrix

$$\mathfrak{M} = \mathfrak{M}(\mathbf{X}) = \left[\frac{\partial S_i}{\partial X_j} \right]. \qquad (5.1.1)$$

If $\mathbf{x}$ is a point on the surface

$$\det(\mathfrak{M}) = \frac{\partial(\mathbf{S})}{\partial(\mathbf{X})} = 0, \qquad (5.1.2)$$

which we call the *Jacobian*, then the linear dependence of the columns of $\mathfrak{M}$ shows that there is a $\mathbf{y} = (y_1, y_2, y_3, y_4) \neq \mathbf{0}$ such that

$$\sum_j y_j \left(\frac{\partial S_i}{\partial X_j} \right) = 0 \qquad (1 \leqslant i \leqslant 4). \qquad (5.1.3)$$

This is just

$$S_i(\mathbf{x}, \mathbf{y}) = 0 \qquad (1 \leqslant i \leqslant 4), \qquad (5.1.4)$$

where $S_i(\ ,\ )$ is the symmetric bilinear form associated to S_i.

A necessary and sufficient condition that $\mathbf{x}$ be on the Jacobian is that there exist a $\mathbf{y}$ with (4). This sets up an involution

$$\mathbf{x} \leftrightarrow \mathbf{y} \qquad (5.1.5)$$

on the Jacobian.

For $\mathbf{x}$ on the Jacobian, we may look alternatively at the rows of the singular matrix $\mathfrak{M}(\mathbf{x})$. A necessary and sufficient condition that $\mathfrak{M}(\mathbf{x})$ be singular is that there is a $\mathbf{z} = (z_1, z_2, z_3, z_4) \neq \mathbf{0}$ such that

$$\left(\sum_i z_i \frac{\partial S_i}{\partial X_j} \right)_{\mathbf{x}} = 0 \qquad (1 \leqslant j \leqslant 4), \qquad (5.1.6)$$

¶ We use a capital J in this context to distinguish from the jacobian $J(\mathcal{C})$.

that is,

$$\left(\frac{\partial S}{\partial X_j}\right)_{\mathbf{x}} = 0 \qquad (1 \leqslant j \leqslant 4), \tag{5.1.7}$$

where

$$S = S_{\mathbf{x}}(\mathbf{X}) = \sum_i z_i S_i(\mathbf{X}). \tag{5.1.8}$$

It follows that the variety $S_{\mathbf{x}} = 0$ is a cone with $\mathbf{x}$ as vertex.¶ The locus of $\mathbf{z}$ is called the *symmetroid* of the four quadrics. Eliminating $\mathbf{x}$ from (7), we see that the equation of the symmetroid is

$$\det\left(\sum_i z_i S_i\right) = 0. \tag{5.1.9}$$

Further, the above sets up a birational correspondence between the symmetroid and the locus of $\mathbf{x}$, i.e. the Jacobian.

2   A special case. In this language, we showed in Chapter 4, Section 3 that the dual $\mathcal{K}^*$ of the Kummer is the symmetroid of the four quadrics

$$\begin{aligned}
S_1: \quad & X_1X_3 - X_2^2, \\
S_2: \quad & X_1X_4 - X_2X_3, \\
S_3: \quad & X_2X_4 - X_3^2, \\
S_4: \quad & f_0X_1^2 + f_1X_1X_2 + f_2X_2^2 + f_3X_2X_3 + f_4X_3^2 + f_5X_3X_4 + f_6X_4^2.
\end{aligned} \tag{5.2.1}$$

We proceed to find the Jacobian.

The intersection Γ of the three quadric surfaces $S_1 = S_2 = S_3 = 0$ is the twisted cubic

$$\Gamma: \qquad (\phi^3, \phi^2\psi, \phi\psi^2, \psi^3), \tag{5.2.2}$$

where (ϕ, ψ) are homogeneous coordinates. A chord is a line joining two points of Γ. By elementary geometry† there is precisely one chord through a given point $\mathbf{x}$ not on Γ, i.e.

$$\mathbf{x} = \lambda\mathbf{m} - \mu\mathbf{l}, \tag{5.2.3}$$

where $\mathbf{l}, \mathbf{m}$ are on Γ. Then

$$\begin{aligned}
S(\lambda\mathbf{m} - \mu\mathbf{l}, \lambda\mathbf{m} + \mu\mathbf{l}) &= \lambda^2 S(\mathbf{m}) - \mu^2 S(\mathbf{l}) \\
&= 0 - 0 = 0
\end{aligned} \tag{5.2.4}$$

¶ i.e. the quadric surface $S_{\mathbf{x}}(\mathbf{X}) = 0$ has a single singularity, namely at $\mathbf{X} = \mathbf{x}$.

† Consider the generators of the quadrics through Γ and $\mathbf{x}$.

for any quadric S through Γ: that is

$$\mathbf{y} = \lambda\mathbf{m} + \mu\mathbf{l} \qquad (5.2.5)$$

has

$$S_j(\mathbf{x}, \mathbf{y}) = 0 \qquad (j = 1, 2, 3). \qquad (5.2.6)$$

A point $\mathbf{x}$ is on the Jacobian of S_1, S_2, S_3, S_4, in the sense of the previous section, if there is a $\mathbf{y}$ such that

$$S_j(\mathbf{x}, \mathbf{y}) = 0 \qquad (j = 1, 2, 3, 4). \qquad (5.2.7)$$

As we saw above, the first three of these conditions, at least for general $\mathbf{x}$, are equivalent to the existence of λ, μ, $\mathbf{l}$, $\mathbf{m}$ such that (3), (5) hold, where without loss of (much) generality we may suppose that

$$\begin{aligned} \mathbf{l} &= (1, \psi, \psi^2, \psi^3), \\ \mathbf{m} &= (1, \omega, \omega^2, \omega^3), \end{aligned} \qquad (5.2.8)$$

for some ψ, ω. As in (4), the last condition in (7) is $\lambda^2 S_4(\mathbf{m}) = \mu^2 S_4(\mathbf{l})$, that is

$$\frac{F(\psi)}{\lambda^2} = \frac{F(\omega)}{\mu^2}. \qquad (5.2.9)$$

3   We now translate into the standard notation of Chapter 2, where $\mathfrak{X} = \{(x, y), (u, v)\}$ is a divisor on $\mathcal{C}: \ Y^2 = F(X)$. Make the substitutions $\psi \mapsto x$, $\lambda \mapsto y$, $\omega \mapsto u$, $\mu \mapsto v$ in (2.3), (2.8) and divide by $x - u$. It follows that the locus of the point

$$\alpha = (\alpha_0, \alpha_1, \alpha_2, \alpha_3), \qquad (5.3.1)$$

where

$$\alpha_j = \frac{u^j y - x^j v}{x - u}, \qquad (5.3.2)$$

is the Jacobian (in the sense of this chapter) of the four quadrics (2.1). This is the promised identification of Weddle's surface with the Jacobian. The six common zeros of the S_j (the base points of the family of quadric surfaces) go into the six nodes of Weddle's surface.

Substituting $\mathbf{X} = \alpha$ in the equation (1.2) of the Jacobian, we get the equation of Weddle's surface:

$$
\begin{aligned}
& f_0(-2\alpha_0{}^3\alpha_3 - 4\alpha_1{}^3\alpha_0 + 6\alpha_0{}^2\alpha_1\alpha_2) \\
& + f_1(-2\alpha_0{}^2\alpha_1\alpha_3 - 2\alpha_1{}^4 + 2\alpha_0\alpha_1{}^2\alpha_2 + 2\alpha_0{}^2\alpha_2{}^2) \\
& + f_2(-2\alpha_0\alpha_1{}^2\alpha_3 - 2\alpha_1{}^3\alpha_2 + 4\alpha_0\alpha_1\alpha_2{}^2) \\
& + f_3(-2\alpha_1{}^3\alpha_3 + 2\alpha_0\alpha_2{}^3) \\
& + f_4(-4\alpha_1{}^2\alpha_2\alpha_3 + 2\alpha_0\alpha_2{}^2\alpha_3 + 2\alpha_1\alpha_2{}^3) \\
& + f_5(-2\alpha_1{}^2\alpha_3{}^2 + 2\alpha_0\alpha_2\alpha_3{}^2 - 2\alpha_1\alpha_2{}^2\alpha_3 + 2\alpha_2{}^4) \\
& + f_6(2\alpha_0\alpha_3{}^3 - 6\alpha_1\alpha_2\alpha_3{}^2 + 4\alpha_2{}^3\alpha_3) \\
& = 0.
\end{aligned}
\tag{5.3.3}
$$

4   Duality. We have already seen in Chapter 4, Section 3 that the symmetroid of the four S_j of the previous sections is the dual of the Kummer. Here we shall give another proof which will also make clear the relation to Weddle's surface. To break a symmetry which may obscure the argument, we start with a bit of Brechtian *Entfremdung* and consider a general $n \times n \times n$ array

$$
s_{ijk} \qquad (1 \leqslant i, j, k \leqslant n)
\tag{5.4.1}
$$

with elements in the ground field. Associated with it are three projective determinantal varieties.

$\mathcal{D}_1$. The locus of the $\mathbf{u} = (u_1, \ldots, u_n)$ such that the $n \times n$ matrix

$$
\sum_i u_i s_{ijk}
\tag{5.4.2}
$$

is singular.

$\mathcal{D}_2$. The locus of the $\mathbf{v}$ with singular $\sum_j v_j s_{ijk}$.

$\mathcal{D}_3$. The locus of the $\mathbf{w}$ with singular $\sum_k w_k s_{ijk}$.

There are birational correspondences between the $\mathcal{D}_j$. If $\mathbf{u}$ is on $\mathcal{D}_1$, so (2) is singular, there is a $\mathbf{v}$ such that

$$
\sum_{i,j} u_i v_j s_{ijk} = 0 \qquad (1 \leqslant k \leqslant n).
\tag{5.4.3}
$$

Hence $\mathbf{v}$ is on $\mathcal{D}_2$, and (3) is a birational correspondence. In this way, we get a triangle

$$\begin{array}{ccc} & \mathcal{D}_1 & \\ \nearrow & & \searrow \\ \mathcal{D}_2 & \longleftrightarrow & \mathcal{D}_3 \end{array} \qquad (5.4.4)$$

of birational correspondences.¶

We want now to find the duals of the $\mathcal{D}_j$, or at least to find the tangent planes. Let $\mathbf{u}, \mathbf{v}$ be a pair of corresponding points under $\mathcal{D}_1 \leftrightarrow \mathcal{D}_2$ and let $\mathbf{u} + d\mathbf{u}$, $\mathbf{v} + d\mathbf{v}$ be an infinitely near pair. Then by (3)

$$\sum_{i,j} u_i dv_j s_{ijk} + \sum_{i,j} du_i v_j s_{ijk} = 0 \qquad (1 \leqslant k \leqslant n). \qquad (5.4.5)$$

Now let $\mathbf{w}$ correspond to $\mathbf{u}$ under $\mathcal{D}_1 \leftrightarrow \mathcal{D}_3$, that is $\sum_{i,k} u_i w_k s_{ijk} = 0$. On multiplying (5) by w_k and summing, we get

$$\sum_{i,j,k} du_i v_j w_k s_{ijk} = 0. \qquad (5.4.6)$$

To sum up: The tangent hyperplane $\mathbf{U}$ at $\mathbf{u}$ has coordinates

$$U_i = \sum_{j,k} v_j w_k s_{ijk}, \qquad (5.4.7)$$

where $\mathbf{v}, \mathbf{w}$ correspond to $\mathbf{u}$.

5   Return to the situation of Section 2, so $n = 4$. Let the coefficients of S_i be s_{ijk}, so

$$s_{ijk} = s_{ikj}. \qquad (5.5.1)$$

In the language of Section 1, the symmetroid is $\mathcal{D}_1$ and $\mathcal{D}_2$, $\mathcal{D}_3$ are both the Jacobian; the $\mathbf{v}, \mathbf{w}$ corresponding to a given $\mathbf{u}$ on the symmetroid are the same.

Now specialize to the four quadrics (2.1) giving $\mathcal{K}^*$, and put

$$w_j = v_j = \alpha_{j-1} = \frac{u^{j-1}y - x^{j-1}v}{x - u} \qquad (5.5.2)$$

¶ It will be seen that it is in general not commutative.

in our standard notation. The formula (4.7) now becomes:

$$U_i = S_i(\alpha). \qquad (5.5.3)$$

It is straightforward to evaluate:

$$\begin{aligned}
(x-u)^2 U_1 &= -(x-u)^2 yv, \\
(x-u)^2 U_2 &= -(x-u)^2 (x+u)yv \\
(x-u)^2 U_3 &= -(x-u)^2 (xu)yv, \\
(x-u)^2 U_4 &= 2y^2 v^2 - F_0(x,u)yv,
\end{aligned} \qquad (5.5.4)$$

so

$$U_1 : U_2 : U_3 : U_4 \;=\; \sigma_0 : \sigma_1 : \sigma_2 : \beta_0. \qquad (5.5.5)$$

This identifies the dual of the symmetroid with the standard Kummer $\mathcal{K}$, as claimed.

Chapter 6

$\mathfrak{G}/2\mathfrak{G}$

0   Introduction. The *Mordell-Weil group* $\mathfrak{G}$ of an abelian variety $\mathcal{A}$ defined over a field k is the group of points of $\mathcal{A}$ defined over k. We shall be concerned with the case when $\mathcal{A}$ is the jacobian of a curve $\mathcal{C}$ defined over k, and then an equivalent definition of $\mathfrak{G}$ is as the group of divisor classes of degree 0 on $\mathcal{C}$ defined¶ over k. If necessary, the ground field under consideration will be denoted by a subscript, e.g. $\mathfrak{G}_k$.

A decisive tool in the investigation of the Mordell-Weil group $\mathfrak{G}$ of a curve of genus 1 is a homomorphism with kernel $2\mathfrak{G}$ into an easily treated group. It generalizes to abelian varieties of higher dimension, but existing versions are not well adapted to the explicit treatment of special cases. In this chapter, we give a version for the jacobian of a curve of genus 2 which can be so used. It is noteworthy that it brings in not just the Kummer but its dual.

We first set up and treat the homomorphism by a simple-minded generalization of an elementary version of the genus 1 homomorphism. Next, we see that it has a natural interpretation in terms of the Kummer, and that it essentially does not distinguish between the different curves $\mathcal{C}_d : dY^2 = F(X)$, where $d \in k^*$. Further, essentially only six of the tropes are used. The need to separate the $\mathcal{C}_d$ leads to the jacobian as a cover of the Kummer, which we started to look at in Chapter 3, Section 8. This brings in the remaining ten tropes in a minor way.

The systematic use of the homomorphism to treat special cases is left to Chapter 11, but we display a local-global phenomenon which has no analogue for genus 1.

We conclude the chapter by considering the curves $Y^2 = F(X)$ with $F(X)$ of degree 5, which are substantially simpler than the general case. Schaefer (1995) gives a generalization to all hyperelliptic curves $Y^2 = F(X)$ with $F(X)$ of odd degree and determines the rank of a curve of genus 3.

¶ For genus 2, see the discussion in Chapter 1 after (1.2.2).

1   The homomorphism. Classical treatments of elliptic curves suggest adjoining the roots of $F(T)$. It turns out to be much more effective to look at all the roots simultaneously.¶ In general $F(T) \in k[T]$ is reducible, and so we introduce the commutative algebra $k[\Theta] = k[T]/F[T]$, where Θ is the image of T. Since $F(T)$ has no multiple factors, by hypothesis, $k[\Theta]$ is the direct sum of fields, one for each irreducible factor of $F(T)$. An element of $k[\Theta]$ is invertible if its image in each of the fields is nonzero: otherwise it is a divisor of zero. The invertible elements form a group under multiplication, which we denote by $k[\Theta]^*$. Any element Λ has a unique representation $\Lambda = L(\Theta)$, where $L[T] \in k[T]$ has degree < 6: it is invertible precisely when the polynomials $L(T)$ and $F(T)$ are coprime.

We use the description of 𝕭 in terms of the effective divisors of degree 2 defined over k which was introduced in Chapter 1. Let

$$\mathfrak{A}_j = \{(a_j, b_j), (c_j, d_j)\} \qquad (j = 1, 2, 3) \tag{6.1.1}$$

be three elements of 𝕭 such that

$$\mathfrak{A}_1 + \mathfrak{A}_2 + \mathfrak{A}_3 = \mathfrak{O} \tag{6.1.2}$$

but otherwise in general position. Then there is an $M(X) \in k[X]$ of degree at most 3 such that the points of the $\mathfrak{A}_j$ are the intersections of $\mathcal{C}$ with $Y = M(X)$. In particular,

$$M(X)^2 - F(X) = e \prod_{j=1}^{3} G_j(X), \tag{6.1.3}$$

where

$$G_j(X) = (X - a_j)(X - c_j) \in k[X] \tag{6.1.4}$$

and† $e \in k^*$. It follows from (3) that

$$\prod_{j=1}^{3} G_j(\Theta) \in k^*\{k[\Theta]^*\}^2. \tag{6.1.5}$$

This leads to the definition

$$\mathcal{L} = k[\Theta]^*/k^*\{k[\Theta]^*\}^2. \tag{6.1.6}$$

¶ cf. Cassels (1991), pp. 66–74. The treatment in this chapter follows the lines of Cassels (1983), but has a much more satisfactory description of the kernel.

† We denote by k^* the group of nonzero elements of k under multiplication.

This is a group, whose neutral element we shall denote by 1. For any

$$\mathfrak{A} = \{(a, b), (c, d)\} \tag{6.1.7}$$

for which¶

$$F(a) \neq 0, \ F(c) \neq 0 \tag{6.1.8}$$

we put

$$\begin{aligned}\mu(\mathfrak{A}) &= (\Theta - a)(\Theta - c)\\ &= \Theta^2 - (a + c)\Theta + ac \in \mathcal{L}.\end{aligned} \tag{6.1.9}$$

The map μ is not always defined, and we must first extend† the definition to the whole of $\mathfrak{G}$. As already noted, $k[\Theta]$ is a direct sum of fields, $\bigoplus K_j$, say. For $\Lambda \in k[\Theta]^*$ let $\Lambda = \bigoplus \Lambda_j$ be the corresponding decomposition. Clearly the image of Λ in $\mathcal{L}$ is unchanged on replacing Λ_j by $\Lambda_j \xi_j^2$ for any $\xi_j \in K_j^*$. Suppose, first, that $a = \theta \in k$ is a root of $F(T)$, so $\Theta \mapsto \theta$ maps $k[\Theta]$ into a component, K_1 say, isomorphic to k. Adjoin a generic point $\mathfrak{x} = (x, y)$. In the (formal) neighbourhood of $\mathfrak{a} = (\theta, 0)$ a local uniformizer is y, and $(x - \theta)F'(\theta)$ is approximately y^2. Hence in the definition (9) of μ one should replace the K_1-component of $(\Theta - a)$ by $-F'(\Theta)$ [note the sign].

The only other case of non-invertibility is when $a = \theta$ is of degree 2 over k and $F(\theta) = 0$. Then $F(T) = (T - \theta)(T - \theta')F_2(T)$, where $F_2(T) \in k[T]$ and $c = \theta'$ in (9). Now $\Theta \mapsto \theta$ maps $k[\Theta]$ into a component K_2 which is a quadratic extension of k. Here we introduce a generic point $\mathfrak{x}$ over $k(\theta)$ and its conjugate $\mathfrak{x}'$ over k. As before, a local uniformizer at $x = \theta$ is y and $(a - \theta)(\theta - \theta')F_2(\theta)$ is approximately y^2. Further, $c - \theta$ is approximately $\theta' - \theta$. Hence the K_2-component of $\Theta^2 - (a + c)\Theta + ac$ in (9) should be replaced by $-F_2(\Theta)$ [note the sign].

We must also deal with the possibility that one or both of the points of $\mathfrak{A}$ is at infinity. If $\mathfrak{A}$ is the intersection of $\mathcal{C}$ with the line at infinity, then $\mathfrak{A} \in \mathfrak{O}$, and clearly $\mu(\mathfrak{X}) = 1$. Otherwise, the points at infinity are defined over k, so $f_6 \in k^{*2}$. Let $\mathfrak{A} = \{\mathfrak{a}, \mathfrak{c}\}$, where $\mathfrak{a}$ is a point at infinity defined over k, and let $\mathfrak{x} = (x, y)$ be a generic point. Then x^{-1} is a local uniformizer at $\mathfrak{a}$, and it follows that $(a - \Theta)$ should be replaced by 1.

The proof above that (2) implies (4) certainly holds when the $\mathfrak{A}$ are generic. With the extended definition of μ it follows that

$$\mu(\mathfrak{A}_1)\mu(\mathfrak{A}_2)\mu(\mathfrak{A}_3) = 1 \in \mathcal{L} \tag{6.1.10}$$

¶ Recall that $\mathrm{Norm}(a - \Theta) = F(a)$.

† There is an alternative treatment in Section 6.

whenever¶
$$\mathfrak{A}_1 + \mathfrak{A}_2 + \mathfrak{A}_3 = \mathfrak{O} \qquad (\mathfrak{A}_j \in \mathfrak{G}). \qquad (6.1.11)$$

In fact the image of μ lies in a smaller group than $\mathcal{L}$. With the notation of (7) we have

$$\text{Norm}_{k[\Theta]/k}(\Theta^2 - (a+c)\Theta + ac)$$
$$= \text{Norm}_{\bar{k}[\Theta]/\bar{k}}(a - \Theta)\text{Norm}_{\bar{k}[\Theta]/\bar{k}}(c - \Theta) \qquad (6.1.12)$$
$$= b^2 d^2.$$

Denote by $\mathcal{M}$ the image in $\mathcal{L}$ of those $\Lambda \in k[\Theta]^*$ for which $\text{Norm}_{k[\Theta]/k}(\Lambda)$ lie in $(k^*)^2$, so $\mathcal{M}$ is a group under multiplication. By (12) we have $\mu(\mathfrak{A}) \in \mathcal{M}$ if (8) holds, and it is clear that this continues to hold for the $\mathfrak{A}$ for which we have had to make special provision.

To sum up:

Lemma 6.1.1. *The map*

$$\mu : \ \mathfrak{G} \to \mathcal{M}, \ \mathfrak{X} \longmapsto \mu(\mathfrak{X}) \qquad (6.1.13)$$

is a group homomorphism. It is independent of the choice of coordinates on $\mathcal{C}$.

All that remains is to prove the last statement. A change of coordinate system consists of a fractional linear transformation of X and the appropriate change on Y [see (1.1.3)]. If Θ' in the new system corresponds to Θ in the old, we have
$$\Theta = (a\Theta' + b)/(c\Theta' + d), \qquad (6.1.14)$$
for a, b, c, d in the ground field. On substituting in (9), the denominator $(c\Theta' + d)^2$ is absorbed in the $k[\Theta]^{*2}$ of (6).

2   The kernel. Since $\mathcal{L}$ has exponent 2, the kernel of μ contains $2\mathfrak{G}$. We shall find that it can be bigger, but not much.

Let $\mathfrak{A}$ be in the kernel. By a preliminary choice of the coordinate system, we may suppose that the support of $\mathfrak{A}$ is in the finite part of the plane. We may suppose that $\mathfrak{A} \not\in \mathfrak{O}$ and we shall also suppose that $\mathfrak{A}$ is not of the shape $\{\mathfrak{c}, \mathfrak{c}\}$, where $\mathfrak{c}$ is a rational point, since they will be mentioned specially later. We thus have

$$\mu(\mathfrak{A}) = 1, \quad \mathfrak{A} = \{\mathfrak{a}, \mathfrak{c}\} = \{(a,b),(c,d)\} \in \mathfrak{G}, \quad a \neq c. \qquad (6.2.1)$$

¶ The argument is sound, but the queasy should await the alternative treatment in Section 6.

Put

$$G(T) = (T - a)(T - c). \qquad (6.2.2)$$

By the definition of μ we have

$$G(\Theta) = n\rho^2, \qquad n \in k, \ \rho \in k[\Theta]. \qquad (6.2.3)$$

Let

$$\begin{aligned} L(T) &= l_0 + l_1 T + l_2 T^2, \\ M(T) &= m_0 + m_1 T + m_2 T^2 + m_3 T^3, \end{aligned} \qquad (6.2.4)$$

where the $l_j \in k$ and the $m_j \in k$ are chosen so that

$$L(\Theta)\rho = M(\Theta), \qquad L(T) \neq 0. \qquad (6.2.5)$$

This is possible: the demand that the coefficients of Θ^4 and Θ^5 on the right hand side vanish imposes two homogeneous linear conditions on the l_j. On squaring, we get

$$n M(\Theta)^2 = G(\Theta)L(\Theta)^2, \qquad (6.2.6)$$

and so

$$n M(T)^2 - G(T)L(T)^2 = wF(T). \qquad (6.2.7)$$

Here $w \in k$ from considerations of degree, and $w \neq 0$ because $a \neq c$.

Let $k(a) = k(\sqrt{g})$, where $g \in k$ and we put $g = 1$ if $a \in k$. Then $b, c, d \in k(\sqrt{g})$ and c, a and d, b are conjugate pairs over k if $g \neq 1$. Let $T \mapsto a$ in (7). Then $nM(a)^2 = wF(a) = wb^2$, and so

$$w/n \in k^* \bigcap k(\sqrt{g})^{*2} = k^{*2} \bigcup gk^{*2}, \qquad (6.2.8)$$

where the second alternative is omitted if $g = 1$.

Take the first alternative in (8), namely that $w/n \in k^{*2}$. On multiplying $L(T)$, $M(T)$ by the same element of k^*, we may suppose that $w = n$, and so that

$$M(T)^2 - n^{-1}G(T)L(T)^2 = F(T), \qquad (6.2.9)$$

where, on changing the sign of M if need be, we may suppose that $M(c) = d$. Let $\mathfrak{B}$ be given by

$$L(X) = 0, \qquad Y = M(X). \qquad (6.2.10)$$

We now distinguish two cases.

If also $M(a) = b$, which is certainly the case by conjugacy if $g \neq 1$, the curve $Y = M(X)$ meets $\mathcal{C}$ in $\mathfrak{A}$ and $\mathfrak{B}$ taken twice. Hence $\mathfrak{A} \in 2\mathfrak{G}$.

If, however, $M(a) = -b$, then $\{(a, -b), (c, d)\} = 2\mathfrak{B}$ and

$$
\begin{aligned}
\mathfrak{C} = \{\mathfrak{c}, \mathfrak{c}\} \qquad & (\mathfrak{c} = (c, d)) \\
= \mathfrak{A} + &\{(a, -b), (c, d)\} \\
= \mathfrak{A} + &2\mathfrak{B}
\end{aligned}
\tag{6.2.11}
$$

is in the kernel.

Conversely, it is straightforward that every divisor of the type $\mathfrak{C} = \{\mathfrak{c}, \mathfrak{c}\}$, where $\mathfrak{c}$ is a point defined over k, is in the kernel. We shall see later that it is not necessarily in $2\mathfrak{G}$.

3   Now consider the second alternative in (2.8), namely $w/n \in gk^{*2}$ where $g \notin k^{*2}$. Without loss of generality $w = n/g$, so

$$
gM(T)^2 - w^{-1}G(T)L(T)^2 = F(T),
\tag{6.3.1}
$$

where by choice of sign of $M(T)$ we have $d = \sqrt{g}M(c)$. By conjugacy $b = -\sqrt{g}M(a)$. It follows, by considering the intersection of $Y = \sqrt{g}M(X)$ with C, that

$$
\{(a, -b), (c, d)\} + 2\mathfrak{B} = \mathfrak{D}
\tag{6.3.2}
$$

where

$$
\mathfrak{B}: \qquad L(X) = 0, \qquad Y = \sqrt{g}M(X)
\tag{6.3.3}
$$

is defined over $k(\sqrt{g})$. Denote conjugacy in $k(\sqrt{g})$ over k by a prime ($'$). For a point $\mathfrak{x} = (x, y)$ let $\bar{\mathfrak{x}} = (x, -y)$. Put $\mathfrak{a} = (a, b)$. Then $(c, d) = \mathfrak{a}'$. In this notation we have

$$
\mathfrak{A} = \{\mathfrak{a}, \mathfrak{a}'\},
\tag{6.3.4}
$$

and (2) is

$$
\{\bar{\mathfrak{a}}, \mathfrak{a}'\} + 2\mathfrak{B} = \mathfrak{D},
\tag{6.3.5}
$$

where by (3)

$$
\mathfrak{B}' = \overline{\mathfrak{B}}.
\tag{6.3.6}
$$

Consider the divisor

$$
\mathfrak{D} = \bar{\mathfrak{a}} \oplus \mathfrak{B}
\tag{6.3.7}
$$

of degree 3, where $\oplus$ denotes addition of divisors. Then by (6) we have

$$
\mathfrak{D}' = \bar{\mathfrak{a}}' \oplus \overline{\mathfrak{B}}.
\tag{6.3.8}
$$

Now $\mathfrak{D}'$ is linearly equivalent to $\mathfrak{D}$ by (5): so the corresponding element of Pic^3 is defined over k, though $\mathfrak{D}$ itself is not (in general).

To sum up so far: we have

$$\mathfrak{A} \oplus \mathfrak{D} \oplus \mathfrak{D}' = (\mathfrak{a} \oplus \mathfrak{a}') \oplus (\bar{\mathfrak{a}} \oplus \mathfrak{B}) \oplus (\bar{\mathfrak{a}}' \oplus \overline{\mathfrak{B}})$$
$$= (\mathfrak{a} \oplus \bar{\mathfrak{a}}) \oplus (\mathfrak{a}' \oplus \bar{\mathfrak{a}}') \oplus (\mathfrak{B} \oplus \mathfrak{B}') \qquad (6.3.9)$$
$$\in 4\mathfrak{O}$$

is in 4 times the canonical class, where $\mathfrak{D}$ is a divisor of degree 3 defined over a quadratic extension of k which is linearly equivalent to its conjugate $\mathfrak{D}'$.

4   We check now that any divisor of the form (3.9) is in the kernel of μ. It is convenient to have a definition:

Definition 6.4.1. *Let $\mathfrak{W}$ denote the set of elements of $\mathfrak{G}$ which are represented by effective divisors $\mathfrak{A}$ with the following property. There is an effective divisor $\mathfrak{D}$ of degree 3 which is either defined over k or defined over a quadratic extension k_2 of k and linearly equivalent to its conjugate $\mathfrak{D}'$ and such that*

$$\mathfrak{A} \oplus \mathfrak{D} \oplus \mathfrak{D}' \in 4\mathfrak{O}. \qquad (6.4.1)$$

First some trivial comments. If $\mathfrak{D}$ is as in the definition and $\mathfrak{A}$ satisfies (1) then $\mathfrak{A}$ is defined over k because it is uniquely given by its linear equivalence class. The divisors $\mathfrak{C} = \{\mathfrak{c}, \mathfrak{c}\}$ of the previous section are in $\mathfrak{W}$, since we can take $\mathfrak{D} = \bar{\mathfrak{c}} + \mathfrak{O}_0$ (say), where $\bar{\mathfrak{c}}$ is the conjugate under $Y \mapsto (-Y)$ and $\mathfrak{O}_0$ is the divisor $X = 0$. Clearly the difference between any two elements of $\mathfrak{W}$ is in $2\mathfrak{G}$. We shall later give a criterion to decide whether or not $\mathfrak{W}$ is in $2\mathfrak{G}$.

Lemma 6.4.1. *The kernel of μ consists precisely of the union of $2\mathfrak{G}$ and $\mathfrak{W}$.*

All that remains to be shown is that $\mathfrak{W}$ is in the kernel. Let $\mathfrak{A} \in \mathfrak{W}$. If $\mathfrak{A} = \{\mathfrak{c}, \mathfrak{c}\}$, there is nothing to prove. Hence we may suppose that $\mathfrak{A}$ is given by (1), where the X-coordinates of the points of the support of $\mathfrak{D}$ are distinct.

Suppose, first, that $\mathfrak{D}$ is not defined over k. As in Chapter 4, Section 2, let $U(X) \in k_2[X]$ of degree 3 have the X-coordinates of the support of $\mathfrak{D}$ as its zeros. Then there is a $V(X) \in k_2[X]$ of degree at most 3 such that

$$\frac{Y + V(X)}{U(X)} \qquad (6.4.2)$$

has divisor $\mathfrak{D}' - \mathfrak{D}$. Its conjugate has divisor $\mathfrak{D} - \mathfrak{D}'$, and so

$$\frac{Y + V(X)}{U(X)} \frac{Y + V'(X)}{U'(X)} = e, \text{ say, } \in k^*, \qquad (6.4.3)$$

that is

$$F(X) + (V(X) + V'(X))Y + V(X)V'(X) = eU(X)U'(X). \qquad (6.4.4)$$

Hence $V(X) + V'(X) = 0$, so $V(X) = \sqrt{g}\,V_0(X)$, where $k_2 = k(\sqrt{g})$, $V_0(X) \in k[X]$, and

$$F[X] = eU(X)U'(X) + gV_0(X)^2. \qquad (6.4.5)$$

Now consider functions on $\mathcal{C}$ of the shape

$$I(X)Y + J(X), \qquad I(X),\, J(X) \in k[X], \qquad (6.4.6)$$

where I, J have respective degrees 1 and 4. There are $2 + 5 = 7$ coefficients, so they can be chosen so that (6) is not zero but vanishes on the divisor $\mathfrak{D} \oplus \mathfrak{D}'$, which is of degree 6 and defined over k. The residual divisor is of degree 2, say $\mathfrak{A} = \{(a,b),(c,d)\}$. On eliminating Y between (6) and the equation of $\mathcal{C}$, we have

$$I(X)^2 F(X) - J(X)^2 = hU(X)U'(X)W(X), \quad h \in k^*,$$
$$W(X) = (X - a)(X - c). \qquad (6.4.7)$$

On substituting $X \mapsto \Theta$ in (5) and (7), we deduce that $W(\Theta) = n\rho^2$, $n \in k$, $\rho \in k[\Theta]$, as required.

The conclusion continues to hold when $\mathfrak{D}$ is defined over k: for then $V'(X) = V(X) \in k[X]$ and we need invoke only (7). This concludes the proof of the theorem.

We conclude with a speculation on a possible more succinct formulation. The elements of Pic of even degree contain a divisor defined over k: for by adding a suitable multiple of the canonical divisor we need consider only Pic^2, when there is a unique effective divisor in the linear equivalence class. We do not know whether elements of Pic of odd degree always contain a divisor defined over at worst a quadratic extension. If so, one could describe the kernel more elegantly as the elements of $J(\mathcal{C}) = \text{Pic}^0$ which lie in the subgroup of Pic generated by 2Pic and the canonical divisor.

5   When is $\mathfrak{W} \subset 2\mathfrak{G}$?

Lemma 6.5.1. *Suppose that $\mathfrak{W}$ is not empty. Then $\mathfrak{W} \subset 2\mathfrak{G}$ precisely when at least one of the following holds:*
(i) *$F(T)$ has a root $\theta \in k$.*
(ii) *The roots of $F(T)$ can be divided into two sets of three roots, where the sets are either defined over k (as wholes) or defined over a quadratic extension and conjugate over k.*

We first show that the conditions (i), (ii) are sufficient. We take the most difficult case and leave the others to the reader.¶ Suppose that $\{\theta_1, \theta_2, \theta_3\}$ is defined over a quadratic extension k_2 of k and that $\{\theta_4, \theta_5, \theta_6\}$ is its conjugate over k. Let $\mathfrak{D} = \{(\theta_1, 0), (\theta_2, 0), (\theta_3, 0)\}$, so that its conjugate $\mathfrak{D}' = \{(\theta_4, 0), (\theta_5, 0), (\theta_6, 0)\}$. Then $\mathfrak{D} \oplus \mathfrak{D}'$ is the divisor of zeros of Y, so $\mathfrak{D} \oplus \mathfrak{D}' \in 3\mathfrak{D}$. But $2\mathfrak{D} = 2(\theta_1, 0) \oplus 2(\theta_2, 0) \oplus 2(\theta_3, 0) \in 3\mathfrak{D}$. Thus $\mathfrak{D}$ and $\mathfrak{D}'$ are linearly equivalent, and their equivalence class is defined over k. Hence $\mathfrak{D} \in \mathfrak{W}$ by Definition 4.1. But the difference between any two elements of $\mathfrak{W}$ is in $2\mathfrak{G}$, so $\mathfrak{W} \subset 2\mathfrak{G}$, as required.

We now proceed in the opposite direction. Suppose that $\mathfrak{W} \cap 2\mathfrak{G}$ is not empty. Then it contains $\mathfrak{D}$. Hence there is an element of Pic^3 defined over k, twice which is the class of $3\mathfrak{D}$. But, as in Chapter 4, the only elements of Pic^3 with the latter property are those containing $(\theta, 0) \oplus \mathfrak{D}$ or $(\theta_i, 0) \oplus (\theta_j, 0) \oplus (\theta_k, 0)$. Hence at least one of these divisor classes must be defined over k. We have shown that the conditions of the Lemma are necessary.

An example may be suggestive. Consider

$$
\begin{aligned}
Y^2 &= F(X) \\
&= C^2 - mQ^2
\end{aligned}
\tag{6.5.1}
$$

where $C = X^3 + \ldots$ is a cubic, Q is a quadratic, and the constant m is not a square. We can arrange that F is irreducible. Let $\mathfrak{D} = \{(\theta_1, 0), (\theta_2, 0), (\theta_3, 0)\}$, where $\theta_1, \theta_2, \theta_3$ are the roots of $C + \sqrt{m}\,Q = 0$. The conjugate of $\mathfrak{D}$ is $\{(\theta_4, 0), (\theta_5, 0), (\theta_6, 0)\}$, so $\mathfrak{D}$ is not rational, but its linear equivalence class is. Let $\mathfrak{a}$ be the point 'at infinity' with $Y/X^3 = 1$. The curve $Y = C + \sqrt{m}\,Q$ meets C in $\mathfrak{a}$ and $\mathfrak{D}$. The residual intersection is the divisor $\mathfrak{B}$ of dimension 2 given by $Q = 0$, $Y = C$. In particular, the equivalence class of $\mathfrak{D}$ contains $\mathfrak{a} \oplus \mathfrak{B}$, which is rational. Finally, $Y = C$ meets C in $\mathfrak{a}$ and $\mathfrak{B}$, both with multiplicity 2: and so $\{\mathfrak{a}, \mathfrak{a}\} \in \mathfrak{W} \cap 2\mathfrak{G}$.

¶ More detail in Flynn, Poonen & Schaefer (1995).

There is another way of looking at the lemma. If¶

$$G(\Theta) = n\rho^2 \qquad\qquad (6.5.2)$$

has a solution, it need not be unique. If (i) of the Lemma holds, we have $F(T) = (T - \theta)H(T)$, where θ is rational. There is then a $\rho_1 \in k[\Theta]$ which satisfies

$$\begin{aligned} \rho_1 &\equiv -\rho \bmod (\Theta - \theta), \\ \rho_1 &\equiv \rho \bmod H(\Theta). \end{aligned} \qquad\qquad (6.5.3)$$

Again, if (ii) holds, there are rational polynomials $R(T)$, $S(T)$ of degree at most 3 and $l, m \in k^*$ such that $F(T) = l\big(R(T)^2 - mS(T)^2\big)$. Then $R(\Theta)^2 = mS(\Theta)^2$, so there is a solution of $\sigma^2 = m$, $\sigma \in k[\Theta]$, and $G(\Theta) = nm^{-1}(\rho\sigma)^2$.

The solution of (2.3) is also not unique when $F(T)$ has a rational factor of degree 2, but the change analogous to (3) does not interchange the two cases in Section 2. Instead it adds the corresponding rational divisor of order 2 to the point obtained.

We do not justify these statements here. They will be apparent after Section 8.

In general, it is easy to decide whether Criterion (i) of Lemma 6.5.1 is satisfied. There is a useful method for deciding whether the more subtle Criterion (ii) is satisfied; namely, to construct the polynomial

$$h(X) = \prod(X - \theta_i\theta_j\theta_k - \theta_\ell\theta_m\theta_n), \qquad\qquad (6.5.4)$$

where the product is taken over the ten unordered partitions of the six roots $\theta_1, \ldots, \theta_6$ of $F(X)$ into two sets of three. Then $h(X)$ is of degree 10 in X and each coefficient is a polynomial in $\theta_1, \ldots, \theta_6$, invariant under the action of S_6. Therefore, $h(X)$ is defined over k, and its coefficients can all be expressed as members of $\mathbb{Z}[f_0/f_6, \ldots, f_5/f_6]$. Clearly Criterion (ii) of Lemma 6.5.1 is satisfied exactly when $h(X)$ has a root in k, which is easy to check. We have provided a file, available by anonymous ftp [the file **kernel.of.mu** in the directory **maple**, as described in Appendix II], which computes $h(X)$ from $F(X)$. The first such use of $h(X)$ was in Flynn, Poonen & Schaefer (1995). We shall also give an example in Chapter 11, Section 5.

¶ We repeat (2.3) for convenience.

6   The Kummer viewpoint. The Kummer interpretation of what we have been doing is suggestive. Recall that for a generic divisor, given by $\mathfrak{X} = \{(x,y),(u,v)\}$, we started with the form $G(\Theta) = \Theta^2 - (x+u)\Theta + xu$. In terms of the standard basis for the jacobian this is just

$$\sigma_0 \Theta^2 - \sigma_1 \Theta + \sigma_2. \qquad (6.6.1)$$

On substituting a root θ_j for Θ we have just a linear form T_j on the coordinates of the Kummer whose vanishing gives a trope which by abuse of language we shall also call T_j. We are thus precisely in Weil's paradigm.¶ A trope meets the Kummer in a conic taken twice. Hence the divisor of zeros of T_j is twice a non-principal divisor. If we substitute $2\mathfrak{Y}$ for the generic divisor $\mathfrak{X}$, the non-principal divisor becomes a principal divisor in terms of $\mathfrak{Y}$. This is clear on taking generic points in Section 1. For tropes T_{ijk} the corresponding result will be proved in Section 7. [For a classical formulation, see Hudson (1905), Section 103.]

We now have an alternative way of dealing with the difficulty discussed in a rather *ad hoc* way in Section 1 when T_j vanishes at a rational divisor $\mathfrak{A}$. It is enough to replace T_j by $T_j\lambda^2$, where λ is any function on the Kummer whose pole cancels the zero of T_j.

The problem of determining the Mordell-Weil group of a curve leads to the following problem: given an element γ (say) of $k[\Theta]^*$, is there a divisor $\mathfrak{A}$ on $\mathcal{C}$ such that $G(\Theta)$ is in the same coset of $k[\Theta]^*$ modulo $k^*k[\Theta]^{*2}$. For this, it is not enough to work with the Kummer, since the Kummer is the same for all curves $\mathcal{C}_d : F(X) = dY^2$. It is necessary to take into account the expression of the jacobian as a double cover of the Kummer given in Chapter 3, Section 9.

7   This section may be omitted at first reading. It is somewhat inconclusive and is referred to again only towards the end of Chapter 16.

The reader may well have been asking: we have an interpretation of six of the tropes; what about the remaining ten? The answer is that they give little more information: only something about which curve $\mathcal{C}_d$, and that in an inconvenient and possibly incomplete form. In this section, we work over the algebraic closure.

¶ More precisely, Weil works with functions, so the paradigm applies to the ratios of two T_j. For genus 1 consider the divisor of $(X - e_1)$ on the curve $Y^2 = (X - e_1)(X - e_2)(X - e_3)$. But already for genus 1 we meet versions in terms of linear forms on the underlying affine space.

For convenience we recall from Chapter 3 the derivation of the equation of a trope T_{ijk}, say T_{123}. Let¶

$$G(X) = (X - \theta_1)(X - \theta_2)(X - \theta_3) = \sum_j g_j X^j,$$

$$H(X) = (X - \theta_4)(X - \theta_5)(X - \theta_6) = \sum_j h_j X^j. \tag{6.7.1}$$

The equation of T_{123} was obtained by expressing

$$\frac{f_6\{G(x)H(u) + G(u)H(x)\} - 2yv}{(x - u)^2} \tag{6.7.2}$$

in terms of the standard basis, namely

$$f_6(g_2 h_0 + g_0 h_2)\sigma_0 + f_6(g_0 + h_0)\sigma_1 + f_6(g_1 + h_1)\sigma_2 + \beta_0. \tag{6.7.3}$$

By abuse of language it will be convenient to use T_{123} for the expressions (2), (3), so

$$T_{123} = \left\{ \frac{\sqrt{f_6}\sqrt{G(x)}\sqrt{H(u)} - \sqrt{f_6}\sqrt{G(u)}\sqrt{H(x)}}{x - u} \right\}^2 \tag{6.7.4}$$

on adjoining the square roots subject to

$$\sqrt{f_6}\sqrt{G(x)}\sqrt{H(x)} = y \text{ and } \sqrt{f_6}\sqrt{G(x)}\sqrt{H(x)} = v. \tag{6.7.5}$$

With the convention

$$T_j = \theta_j{}^2 - (x + u)\theta_j + xu = (x - \theta_j)(u - \theta_j), \tag{6.7.6}$$

we have
$$T_1 T_2 T_3 = G(x)G(u). \tag{6.7.7}$$

Hence
$$T_1 T_2 T_3 T_{123} = v^2, \tag{6.7.8}$$

where
$$\nu = \frac{G(u)y - G(x)v}{x - u}$$
$$= \sum_j g_j \alpha_j \tag{6.7.9}$$

¶ Not the same G as in the previous section.

in terms of the standard jacobian basis. Hence ν is one of the functions introduced in Chapter 3, Section 9. But it is not, in general, defined over the ground field.

8   **The norm of ρ.** As already remarked, it is not really surprising that the μ invariant does not completely decide on membership of $2\mathfrak{G}$ because it makes no reference at all to the Y-coordinates.¶ In this section, we look at this from another point of view. In Sections 1, 2 we introduced $\rho \in k[\Theta]$. The square of its norm is easily found in terms of the other quantities introduced, but the sign of the norm gives extra information.

Recall the notation and identities of Sections 1, 2 where, as usual, we take $\mathfrak{X} = \{(x, y), (u, v)\}$ and:

$$G(T) = (T - x)(T - u), \tag{6.8.1.a}$$
$$G(\Theta) = n\rho^2, \tag{6.8.1.b}$$
$$L(\Theta)\rho = M(\Theta), \tag{6.8.1.c}$$
$$L(T) = l_2 T^2 + l_1 T + l_0, \tag{6.8.1.d}$$
$$M(T) = m_3 T^3 + m_2 T^2 + m_1 T + m_0, \tag{6.8.1.e}$$
$$M(T)^2 - n^{-1} G(T) L(T)^2 = wF(T). \tag{6.8.1.f}$$

The equation of the Kummer is quadratic in β_0, the two roots corresponding to the choices of sign for yv with $y^2 v^2 = F(x)F(u)$. If Norm denotes the norm from $k[\Theta]$ to k, we have $\text{Norm}(\Theta^2 - (x + u)\Theta + xu) = y^2 v^2$. It is thus reasonable to hope that $\text{Norm}(\rho)$ will give information about the sign of yv, and that is in fact the case:

Lemma 6.8.1.

$$\text{Norm}\,\rho = -(wf_6)^{-1} n^{-3} M(x) M(u). \tag{6.8.2}$$

We need resultants.† Let $P(T)$, $Q(T) \in k[T]$ have precise degrees p, q and top coefficients a, b. We write

$$R(P, Q) = a^q b^p \prod_{\pi, \kappa} (\pi - \kappa)$$
$$= a^q \prod_{\pi} Q(\pi) \tag{6.8.3}$$

¶ Ed Schaefer points out that in the quintic case the μ invariant *does* completely determine membership. We treat the quintic case in Section 10. It will be a useful exercise for the reader to analyse why Ed's remark is true.

† cf. Cassels (1991), Chapter 16

where π, κ run through the roots of $P(T)$, $Q(T)$. Then

$$R(Q, P) = (-1)^{pq} R(P, Q) \text{ and } R(P, Q_1 Q_2) = R(P, Q_1) R(P, Q_2). \quad (6.8.4)$$

Further, $R(P, Q_1) = R(P, Q_2)$ if $Q_1 \equiv Q_2 \mod P$. It will be convenient to use a suffix 0 to denote a polynomial divided by its top coefficient. Let θ run through the roots of F_0.

We have to find

$$\text{Norm}\, \rho = \prod_\theta \rho(\theta) = \frac{\prod_\theta M(\theta)}{\prod_\theta L(\theta)} = \frac{R(F_0, M)}{R(F_0, L)} = \frac{m_3^6 R(F_0, M_0)}{l_2^6 R(L_0, F_0)}. \quad (6.8.5)$$

Now

$$
\begin{aligned}
R(M_0, F_0) &= (w f_6)^{-3} R(M_0, wF) \\
&= (w f_6)^{-3} R(M_0, -n^{-1} G L^2) \qquad \text{by } (1.f) \qquad (6.8.6) \\
&= (w f_6)^{-3} (-n^{-1} l_2^2)^3 R(M_0, G_0) R(M_0, L_0)^2
\end{aligned}
$$

and

$$
\begin{aligned}
R(L_0, F_0) &= (w f_6)^{-2} R(L_0, wF) \\
&= (w f_6)^2 R(L_0, M^2) \qquad \text{by } (1.f) \qquad (6.8.7) \\
&= (w f_6)^{-2} m_3^4 R(L_0, M_0)^2.
\end{aligned}
$$

But $R(L_0, M_0) = R(M_0, L_0)$, so

$$
\begin{aligned}
\text{Norm}\, \rho &= \frac{m_3^6 R(M_0, F_0)}{l_2^6 R(L_0, F_0)} \\
&= -(w f_6)^{-1} n^{-3} m_3^2 R(M_0, G_0) \qquad\qquad\qquad (6.8.8) \\
&= -(w f_6)^{-1} n^{-3} R(M, G_0).
\end{aligned}
$$

But $R(M, G_0) = M(x) M(u)$: and we are done.

Corollary. (1) *gives an expression of* $\mathfrak{X} \in 2\mathfrak{G}$ *if and only if*

$$\text{Norm}\, \rho = -f_6^{-1} n^{-3} yv. \quad (6.8.9)$$

The enunciation requires a little care because, as we saw in Section 5, it is possible that (1) gives $\mathfrak{X} \in \mathfrak{W}$, but there is another representation which gives also $\mathfrak{X} \in 2\mathfrak{G}$.

By adjoining a square root if necessary, we may suppose that $w = 1$ and $M(x) = y$. Then (9) holds if and only if $M(u) = v$, which, as we saw in Section 2, is necessary and sufficient to give $\mathfrak{X} \in 2\mathfrak{G}$.

The statements made without proof at the end of Section 5 follow readily from (9). There is also a minor complement to the results of Section 7. Let ρ_j be the result of substituting the root θ_j of $F(T)$ for Θ in $\rho \in k[\Theta]$. In the notation of Section 7, it is easy to see that $\nu/\rho_1\rho_2\rho_3$ is invariant under any Galois automorphism which permutes $\{\theta_1, \theta_2, \theta_3\}$. It may be deduced from (9) that it is also fixed by any automorphism which takes $\{\theta_1, \theta_2, \theta_3\}$ into $\{\theta_4, \theta_5, \theta_6\}$.

9   A pathology. We construct a curve C of genus 2 and an element $\mathfrak{A}$ of its jacobian, both defined over $\mathbb{Q}$, such that

(i) $\mathfrak{A}$ is not divisible by 2 over $\mathbb{Q}$,
   but
(ii) $\mathfrak{A}$ is divisible by 2 over every $\mathbb{Q}_p$ and over $\mathbb{R} = \mathbb{Q}_\infty$.

It is not difficult to see that the analogous behaviour is impossible in genus 1. We use the fact that over every $\mathbb{Q}_p$ at least one of 2, 17, 34 is a square.

Let $A(X)$, $B(X)$, $C(X)$ in $\mathbb{Q}[X]$ be quadratics with constant term 1. Suppose that all are irreducible over $\mathbb{Q}$, but that they split over $\mathbb{Q}(\sqrt{2})$, $\mathbb{Q}(\sqrt{17})$ and $\mathbb{Q}(\sqrt{34})$ respectively. Consider the curve

$$C : \quad Y^2 = F(X) = A(X)B(X)C(X). \tag{6.9.1}$$

The point $\mathfrak{a} = (1, 1)$ is on C. Put $\mathfrak{A} = \{\mathfrak{a}, \mathfrak{a}\}$. Then $\mathfrak{A} \notin 2\mathfrak{G}$ by Lemma 5.1. On the other hand, for every p, including $p = \infty$, there is a $t_p \in \mathbb{Q}_p$ with $F(t_p) = 0$. Put $\mathfrak{B}_p = \{\mathfrak{a}, (t_p, 0)\}$. Then $\mathfrak{A} = 2\mathfrak{B}_p$.

10   The quintic case. The situation for curves $C : Y^2 = F(X)$ with square free $F(X)$ of degree 5, which we are systematically excluding from consideration, is particularly simple. In this section we reduce it to the general case. Let $k[\Theta] = k[T]/F[T]$, which now is a commutative algebra of degree 5 over the ground field k, and put $\mathcal{L}' = k[\Theta]^*/(k[\Theta]^*)^2$. For a rational divisor $\mathfrak{X} = \{(x, y), (u, v)\}$ with $yv \neq 0$ define $\mu'(\mathfrak{X})$ to be the image in $\mathcal{L}'$ of $(\Theta - x)(\Theta - u) \in k[\Theta]^*$. The extension to $\mathfrak{X}$ with $yv = 0$ is on the same lines as the corresponding extension in Section 1, and is left to the reader.

THEOREM 6.10.1. *The map μ' is a group homomorphism. The kernel is precisely* $2\mho$.

After a substitution $X \mapsto X + \text{constant}$, we may suppose without loss of generality that $F(0) \neq 0$. Put $H(X) = X^5 F(X^{-1})$ and $G(X) = XH(X)$, so G has precise degree 6; the curves defined by $Y^2 = F(X)$ and $Y^2 = G(X)$ are birationally equivalent, using the map $(X, Y) \mapsto (1/X, Y/X^3)$ in either direction. The general theory applies to the curve $\mathcal{D} : Y^2 = G(X)$, which is isomorphic to $\mathcal{C}$. We identify the Mordell-Weil group of $\mathcal{D}$ with $\mho$.

Let $k[\Phi] = k[T]/G(T)$. By Lemma 1.1 applied to $\mathcal{D}$, there is a homomorphism μ from $\mho$ to $\mathcal{L}$ (say) $= k[\Phi]^* / \big(k^*(k[\Phi]^*)^2\big)$. The kernel is precisely $2\mho$ by Lemma 5.1 (i).

$k[\Phi]$ is the direct sum of k and $k[\Theta]$, the maps into the components taking Φ into 0 and Θ^{-1} respectively. Hence $k[\Phi]^*/k^*$ is isomorphic as a group to $k[\Theta]^*$; the isomorphism taking the class of $m \oplus \sigma \in k[\Phi] = k \oplus k[\Theta]$ modulo k^* into $m^{-1}\sigma$. It follows that $\mathcal{L}'$ and $\mathcal{L}$ are isomorphic. Treating the isomorphism as an identification, it is easy to see that the two maps μ' and μ are the same.

Chapter 7

The jacobian over local fields.
Formal groups

0   Introduction. The formal group law mentioned in Chapters 2, 3 helps us to describe properties of the jacobian over a local field. We shall expand on the comment made at the end of Chapter 3, Section 9, that the formal group may be derived from the biquadratic and bilinear forms on the jacobian. We shall show explicitly how to derive terms of the formal group and outline some of its properties, such as the associated invariant differential, logarithm and exponential maps. Some consequences of these properties will prove useful when we perform computations on the Mordell-Weil group in Chapter 11.

1   Computing the formal group. There are many similarities between our $\mathbb{P}^{15}$ embedding of the jacobian and the $\mathbb{P}^3$ embedding of an elliptic curve. We shall first show how the formal group of an elliptic curve may be derived from its $\mathbb{P}^3$ embedding and then outline the analogous computations for the jacobian of a curve of genus 2.

Let $\mathcal{E}$ be an elliptic curve over the ground field k in the shape

$$\mathcal{E}: \quad Y^2 = f_0 + f_1 X + f_2 X^2 + X^3, \qquad (7.1.1)$$

where the cubic in X has no multiple factors. Then the jacobian has an embedding into $\mathbb{P}^3$ given by the coordinates

$$z_0 = X^2, \quad z_1 = Y, \quad z_2 = X, \quad z_3 = 1. \qquad (7.1.2)$$

We shall denote the projective locus of $(z_0, \ldots, z_3)$ in $\mathbb{P}^3$ by $J(\mathcal{E})$. The defining equations for $J(\mathcal{E})$ are given by a pair of quadratic relations

$$z_0 z_2 = z_1^2 - f_0 z_3^2 - f_1 z_2 z_3 - f_2 z_2^2$$
$$z_0 z_3 = z_2^2. \qquad (7.1.3)$$

If we now define the *localized coordinates* to be

$$s_i = z_i/z_0, \quad i = 0, \ldots, 3, \qquad (7.1.4)$$

then the defining equations can be normalized to become

$$s_2 = s_1^2 - f_0 s_3^2 - f_1 s_2 s_3 - f_2 s_2^2$$
$$s_3 = s_2^2. \tag{7.1.5}$$

By recursive substitution, these equations formally define power series

$$\sigma_2(\mathbf{s}), \sigma_3(\mathbf{s}) \in \mathbb{Z}_f[[\mathbf{s}]] \tag{7.1.6}$$

where

$$\mathbf{s} = s_1, \quad \mathbb{Z}_f = \mathbb{Z}[f_0, \dots, f_3]. \tag{7.1.7}$$

These give formal identities $s_i = \sigma_i(\mathbf{s})$, for $i = 2, 3$, which can be extended to $i = 0, 1$ by defining $\sigma_0(\mathbf{s}) = 1$ and $\sigma_1(\mathbf{s}) = s_1$.

There are objects on $J(\mathcal{E})$ analogous to all of those in Chapter 3, and in particular there are biquadratic forms Ψ_{IJ} which satisfy precisely the same conditions as in Theorem 3.9.1. The relevant row here is: Ψ_{0J}, for $J = 0, \dots, 3$. If we denote $a_i = z_i(\mathfrak{A})$ and $b_i = z_i(\mathfrak{B})$, for $i = 0, \dots, 3$, then these biquadratic forms can be written out explicitly as

$$\Psi_{00} = R(\mathbf{a}, \mathbf{b})^2, \qquad \Psi_{01} = b_1 S(\mathbf{a}, \mathbf{b}) + a_1 S(\mathbf{b}, \mathbf{a}),$$
$$\Psi_{02} = R(\mathbf{a}, \mathbf{b}) \cdot T(\mathbf{a}, \mathbf{b}), \qquad \Psi_{03} = T(\mathbf{a}, \mathbf{b})^2, \tag{7.1.8}$$

where

$$
\begin{aligned}
R(\mathbf{a}, \mathbf{b}) =\ & a_0 b_0 - 2 f_1 a_2 b_2 - 4 f_0 a_3 b_2 - 4 f_0 a_2 b_3 + f_1^2 a_3 b_3 - 4 f_2 f_0 a_3 b_3, \\
S(\mathbf{a}, \mathbf{b}) =\ & a_0^2 b_0 + 2 f_2 a_2 a_0 b_0 + 2 f_1 a_2 a_0 b_2 + 4 f_0 a_2 a_0 b_3 \\
& + 3 f_1 a_3 a_0 b_0 + 12 f_0 a_3 a_0 b_2 - 3 f_1^2 a_3 a_0 b_3 \\
& + 12 f_0 f_2 a_3 a_0 b_3 + 4 f_0 a_3 a_2 b_0 - 2 f_1^2 a_3 a_2 b_2 \\
& + 8 f_0 f_2 a_3 a_2 b_2 - 2 f_1^2 f_2 a_3 a_2 b_3 - 4 f_0 f_1 a_3 a_2 b_3 \\
& + 8 f_0 f_2^2 a_3 a_2 b_3 - f_1^3 a_3^2 b_3 - 8 f_0^2 a_3^2 b_3 + 4 f_0 f_1 f_2 a_3^2 b_3, \\
T(\mathbf{a}, \mathbf{b}) =\ & 2 a_1 b_1 + a_2 b_0 + a_0 b_2 + 2 f_2 a_2 b_2 + f_1 a_2 b_3 + f_1 a_3 b_2 + 2 f_0 a_3 b_3.
\end{aligned}
\tag{7.1.9}
$$

Note that this gives a nice description of the group law on an elliptic curve which is non-degenerate at duplication. That is, putting $\mathbf{b} = \mathbf{a}$ in (8) gives the duplication law.

If $\mathfrak{A}, \mathfrak{B}$ have corresponding $s_i = z_i(\mathfrak{A})/z_0(\mathfrak{A})$ and $t_i = z_i(\mathfrak{B})/z_0(\mathfrak{B})$, then we can normalize the Ψ_{IJ}, by dividing through by $a_0^2 b_0^2$, so that they are polynomials in the s_i, t_i. On substituting $s_i = \sigma_i(\mathbf{s})$, $t_i = \sigma_i(\mathbf{t})$, where σ_i are as in (6), the Ψ_{IJ} can then be expressed as power series in $\mathbf{s}, \mathbf{t}$ defined over $\mathbb{Z}_f$. The power series corresponding to Ψ_{00} includes the term 1 and so is invertible in $\mathbb{Z}_f[[\mathbf{s}, \mathbf{t}]]$. We may therefore define the following power series:

$$\mathcal{F}(\mathbf{s}, \mathbf{t}) = \Psi_{01}/\Psi_{00} \in \mathbb{Z}_f[[\mathbf{s}, \mathbf{t}]], \qquad (7.1.10)$$

which formally gives $u = z_1(\mathfrak{A} + \mathfrak{B})/z_0(\mathfrak{A} + \mathfrak{B})$. The above construction shows that all of the information required to derive terms of $\mathcal{F}$ is contained in the defining equations (3) and the biquadratic forms Ψ_{00}, Ψ_{01} in (8). In fact, given the linear terms $\mathcal{F}(\mathbf{s}, \mathbf{t}) = \mathbf{s} + \mathbf{t} + \ldots$, it is sufficient to use the simpler bilinear forms Φ_{ij} to derive $\mathcal{F}(\mathbf{s}, \mathbf{t})^2$, from which $\mathcal{F}(\mathbf{s}, \mathbf{t})$ can then be computed.

Let us return now to the jacobian $J(\mathcal{C})$ of a curve $\mathcal{C}$ of genus 2, embedded into $\mathbb{P}^{15}$ by the coordinates $z_0, \ldots, z_{15}$ defined at the beginning of Chapter 2, Section 2. In this case, there are 72 independent quadratic relations, such as the one given in (2.3.6), which are available by anonymous ftp as explained in Appendix II. These give a set of defining equations for the jacobian. To imitate the above construction, only 13 of these relations are required, namely those which contain the monomial $z_0 z_i$ for $i = 3, \ldots, 15$. The *localized* coordinates are $s_i = z_i/z_0$, as usual. On performing the usual normalization, equivalent here to dividing through by z_0^2, and performing recursive substitution on these 13 equations, we obtain power series

$$\sigma_i(\mathbf{s}) \in \mathbb{Z}_f[[\mathbf{s}]], \quad i = 3, \ldots, 15, \qquad (7.1.11)$$

where

$$\mathbf{s} = \begin{pmatrix} s_1 \\ s_2 \end{pmatrix}, \quad \mathbb{Z}_f = \mathbb{Z}[f_0, \ldots, f_6]. \qquad (7.1.12)$$

These can be formally extended to $i = 0, 1, 2$ by defining $\sigma_0(\mathbf{s}) = 1$ and $\sigma_i(\mathbf{s}) = s_i$ for $i = 1, 2$. We can now define the vector of power series

$$\mathcal{F}(\mathbf{s}, \mathbf{t}) = \begin{pmatrix} \mathcal{F}_1(\mathbf{s}, \mathbf{t}) \\ \mathcal{F}_2(\mathbf{s}, \mathbf{t}) \end{pmatrix} \qquad (7.1.13)$$

where

$$\mathcal{F}_1(\mathbf{s}, \mathbf{t}) = \Psi_{01}/\Psi_{00}, \quad \mathcal{F}_2(\mathbf{s}, \mathbf{t}) = \Psi_{02}/\Psi_{00} \in \mathbb{Z}_f[[\mathbf{s}, \mathbf{t}]]. \qquad (7.1.14)$$

In this situation, we can again, given the linear terms $\mathcal{F}_i(\mathbf{s}, \mathbf{t}) = s_i + t_i + \ldots$, for $i = 1, 2$, make do with the bilinear forms Φ_{ij} of Lemma 3.9.1, which give each $\mathcal{F}_i(\mathbf{s}, \mathbf{t})^2$, from which each $\mathcal{F}_i(\mathbf{s}, \mathbf{t})$ can be derived. Up to cubic terms $\mathcal{F}$ is given by

$$\begin{aligned} \mathcal{F}_1 &= s_1 + t_1 + 2f_4 s_1^2 t_1 + 2f_4 s_1 t_1^2 - f_1 s_2^2 t_2 - f_1 s_2 t_2^2 + \ldots \\ \mathcal{F}_2 &= s_2 + t_2 + 2f_2 s_2^2 t_2 + 2f_2 s_2 t_2^2 - f_5 s_1^2 t_1 - f_5 s_1 t_1^2 + \ldots \end{aligned} \qquad (7.1.15)$$

The lazy reader, who does not wish to perform the above substitutions, may directly access the terms of the formal group up to degree 7 for a general curve of genus 2. We have placed them in a file available by anonymous ftp, described in Appendix II.

So far, these power series are merely formal. We can consider $J(\mathcal{E})$ and $J(\mathcal{C})$ together, which we shall denote J in either case. We say that $\mathbf{s}$ is a *local parameter* for J (that is: s_1 in the elliptic curve case and the pair s_1, s_2¶ in the genus 2 case). Assume that the ground field k is a non-archimedean local field (that is to say, a field complete with respect to a non-archimedean valuation), and that the coefficients f_i satisfy $|f_i| \leqslant 1$. Define a neighbourhood of $\mathfrak{O}$

$$\mathcal{N} = \{\mathfrak{A} \in J(k): \ |s_i(\mathfrak{A})| < 1, \ i > 0\}. \tag{7.1.16}$$

For any $\mathfrak{A}$ in $\mathcal{N}$, the power series σ_i converge, giving that $\mathfrak{A}$ is uniquely defined by its local parameter $\mathbf{s}$. The power series $\mathcal{F}$ also converges and gives the group law on $\mathcal{N}$. The properties of the group law — associativity, commutativity, and so on — therefore induce corresponding properties on the power series $\mathcal{F}$. This is summarized in the following theorem.

THEOREM 7.1.1. *The vector of power series $\mathcal{F}$ satisfies*

(a) $\mathcal{F}(\mathbf{s}, \mathbf{0}) = \mathbf{s}, \quad \mathcal{F}(\mathbf{0}, \mathbf{t}) = \mathbf{t}$,
(b) $\mathcal{F}(\mathcal{F}(\mathbf{s}, \mathbf{t}), \mathbf{u}) = \mathcal{F}(\mathbf{s}, \mathcal{F}(\mathbf{t}, \mathbf{u}))$,
(c) $\mathcal{F}(\mathbf{s}, \mathbf{t}) = \mathcal{F}(\mathbf{t}, \mathbf{s})$.

Suppose that the ground field k is a non-archimedean local field, and the f_i satisfy $|f_i| \leqslant 1$. Let $\mathfrak{A}, \mathfrak{B}, \mathfrak{C} \in \mathcal{N}$ have local parameters $\mathbf{s}, \mathbf{t}, \mathbf{u}$, respectively. Then $\mathbf{u} = \mathcal{F}(\mathbf{s}, \mathbf{t})$ in k.

Further details of the formal group in the genus 2 situation can be found in Flynn (1990a). There is also a discussion in Grant (1990) of the simplified genus 2 situation when the curve $\mathcal{C}$ has a rational Weierstrass point, and so can be written in the form $Y^2 = $ (quintic in X); this allows the use of a $\mathbb{P}^8$ embedding of $J(\mathcal{C})$.

In principle, the above could be imitated for a hyperelliptic curve of genus $g > 2$. There would be an embedding of the jacobian in $\mathbb{P}^{4^g - 1}$, for which the defining relations would be quadratic, and there would be objects analogous to the Φ_{ij}. The computer algebra involved would be prohibitive, and so it would be useful to have a direct method for deriving the terms of $\mathcal{F}$ without needing large intermediate objects such as the Φ_{ij}.

¶ Note that, in the genus 2 case, s_1, s_2 are the same as μ, λ of (2.3.3).

2   **General properties of formal groups.** The properties of the power series $\mathcal{F}$ may be abstracted as axioms. The modern approach, outlined here, is to explore what follows directly from these axioms. The definition will be for the general n-parameter situation, since it is no more difficult to describe than the one- and two-parameter situations.

Definition 7.2.1. *Let* $\mathbf{s} = (s_i)$ *be a column vector of indeterminates* $s_1, \ldots, s_n$, *and similarly* $\mathbf{t}, \mathbf{u}$. *Let* $\mathcal{F}(\mathbf{s}, \mathbf{t}) = \big(\mathcal{F}_i(\mathbf{s}, \mathbf{t})\big)$ *be a vector of* n *power series over a ring* R, *which is an integral domain of characteristic 0. Then* $\mathcal{F}$ *is an* n-*parameter formal group if it satisfies the conditions* $(a), (b), (c)$ *listed in Theorem 1.1.*

The power series in (1.10) induced by $J(\mathcal{E})$ is a one-parameter formal group, and that in (1.13) induced by $J(\mathcal{C})$ is a two-parameter formal group, both of which are defined over the ring $\mathbb{Z}_f$. The same definition of a formal group can also be made for rings of finite characteristic, but only the characteristic 0 situation will be required here.

When θ is any power series in several variables, formally define $\partial\theta/\partial x$ to be the power series obtained be replacing each occurrence of x^n by nx^{n-1}. The symbol d is defined to satisfy the relations

$$\mathrm{d}\big(\theta(x)\big) = (\partial\theta/\partial x) \cdot \mathrm{d}x, \quad \mathrm{d}(\theta\rho) = \theta \cdot \mathrm{d}(\rho) + \rho \cdot \mathrm{d}(\theta). \tag{7.2.1}$$

When $f = (f_i)$ is a vector of n power series in $\mathbf{s}$, its derivative f' is given by the following standard matrix, and similarly for $\mathcal{F}_\mathbf{s}$, $\mathcal{F}_\mathbf{t}$, the partials of the formal group with respect to $\mathbf{s}$ and $\mathbf{t}$.

$$f' = \big(\partial f_i/\partial s_j\big), \quad \mathcal{F}_\mathbf{s} = \big(\partial\mathcal{F}_i/\partial s_j\big), \quad \mathcal{F}_\mathbf{t} = \big(\partial\mathcal{F}_i/\partial t_j\big). \tag{7.2.2}$$

The symbol d can be extended to vectors by letting d**s** denote the vector $(\mathrm{d}s_i)$. This allows the formal chain rule on power series to be written as $\mathrm{d}f = f' \cdot \mathrm{d}\mathbf{s}$. It is now possible to define

$$\omega(\mathbf{t}) = P(\mathbf{t}) \cdot \mathrm{d}\mathbf{t}, \quad P(\mathbf{t}) = \big(P_{ij}(\mathbf{t})\big) = \mathcal{F}_\mathbf{s}(\mathbf{0}, \mathbf{t})^{-1}. \tag{7.2.3}$$

Note that the determinant of the matrix $\mathcal{F}_\mathbf{s}(\mathbf{0}, \mathbf{t})$ is an invertible power series since it has constant term 1; therefore $P(\mathbf{t})$ is also a matrix of power series defined over R. Clearly $P(\mathbf{0})$ is the identity matrix. The proof of the following theorem and its corollary is the same as that for the one-parameter case given in [Silverman (1986), pp. 119–121].

Lemma 7.2.1. *The differential form* ω *of (3) is an invariant differential; that is to say, it satisfies* $\omega \circ \mathcal{F}(\mathbf{t}, \mathbf{s}) = \omega(\mathbf{t})$. *Any invariant differential must be of the form* $a\omega$ *for some* $a \in R$.

Corollary. *Let $[n]$ represent multiplication by n, defined inductively by $[0](\mathbf{t}) = 0$ and $[n+1](\mathbf{t}) = \mathcal{F}([n](\mathbf{t}), \mathbf{t})$. Then $\omega \circ [n]$ is an invariant differential and $\omega \circ [n] = n\omega$, so that $[n]'(T) \in nR[[T]]$. It follows that, for any prime p, there exist $f(\mathbf{t}), g(\mathbf{t})$ defined over R such that $[p](\mathbf{t}) = p \cdot f(\mathbf{t}) + g(\mathbf{t}^p)$, where $\mathbf{t}^p = (t_i^p)$.*

An interesting fact about n-parameter formal groups, not immediately apparent from the axioms, is they are all isomorphic over a field of characteristic 0. This can be described by the following logarithm map which gives a formal isomorphism between an n-parameter formal group and the additive group $\mathbf{s} + \mathbf{t}$.

Definition 7.2.2. *The formal logarithm of $\mathcal{F}$ is $L = (L_i)$, where each L_i is a power series in $\mathbf{s}$ over the field of fractions of R, defined by*

$$L(\mathbf{s}) = \mathbf{s} + \ \text{terms of higher degree}, \quad L'(\mathbf{s}) = P(\mathbf{s}), \qquad (7.2.4)$$

where P is as in (3). Define $E = (E_i)$, the formal exponential map of $\mathcal{F}$, by $E(L(\mathbf{s})) = L(E(\mathbf{s})) = \mathbf{s}$.

The existence and uniqueness of such power series can be shown by induction; see Theorem 1.27 on page 15 of Zink (1984). The justification for the following lemma is the same as for the one-parameter case, described in [Silverman (1986), p. 122].

Lemma 7.2.2. *The formal logarithm and exponential maps satisfy the formal identities $L(\mathcal{F}(\mathbf{s}, \mathbf{t})) = L(\mathbf{s}) + L(\mathbf{t})$ and $E(\mathbf{s} + \mathbf{t}) = \mathcal{F}(E(\mathbf{s}), E(\mathbf{t}))$.*

The power series which occur in L and E are defined over the field of fractions of the ring R. It will be useful to know what denominators occur in the coefficients of the power series. Differentiating both sides of the equation $E(\mathbf{s} + \mathbf{t}) = \mathcal{F}(E(\mathbf{s}), E(\mathbf{t}))$ with respect to $\mathbf{t}$, and then evaluating at $\mathbf{t} = 0$ gives

$$E'(\mathbf{s}) = \mathcal{F}_{\mathbf{t}}(E(\mathbf{s}), 0) \cdot E'(0) = \mathcal{F}_{\mathbf{t}}(E(\mathbf{s}), 0), \qquad (7.2.5)$$

since $E'(0)$ is the identity matrix. Taking derivatives with respect to $\mathbf{s}$ and evaluating at $\mathbf{s} = 0$ gives, by induction on r, that $\partial^r E_k / (\partial s_1^{i_1} \dots \partial s_n^{i_n}) \in R$ and is 0 when r is even. A similar argument shows the same for each L_k. Therefore, each E_k, L_k can be written in the form

$$\sum \big(a_{i_1 \dots i_n} / (i_1! \dots i_n!) \big) s_1^{i_1} \dots s_n^{i_n}, \quad a_{i_1 \dots i_n} \in R. \qquad (7.2.6)$$

This also gives an inductive method of computing the coefficients of E and L. For the two-parameter formal group induced by $J(\mathcal{C})$ in Section 1, the terms up to degree 3 in **s** are as follows.

$$L_1(\mathbf{s}) = s_1 + \frac{1}{3}(-2f_4 s_1^3 + f_1 s_2^3) + \dots$$

$$E_1(\mathbf{s}) = s_1 + \frac{1}{3}(2f_4 s_1^3 - f_1 s_2^3) + \dots$$

$$L_2(\mathbf{s}) = s_2 + \frac{1}{3}(-2f_2 s_2^3 + f_5 s_1^3) + \dots$$

$$E_2(\mathbf{s}) = s_2 + \frac{1}{3}(2f_2 s_2^3 - f_5 s_1^3) + \dots$$

(7.2.7)

3 The reduction map. When k is a non-archimedean local field, with valuation ring $\mathfrak{o}$ and maximal ideal $\mathfrak{p}$,

$$\mathfrak{o} = \{x \in k : \ |x| \leqslant 1\}, \quad \mathfrak{p} = \{x \in k : \ |x| < 1\}, \qquad (7.3.1)$$

then the natural surjection $\ ^\sim : \ \mathfrak{o} \to \mathfrak{o}/\mathfrak{p}$ onto the residue field induces a map on projective space. For any $\mathbf{x} = (x_i) \in \mathbb{P}^n(k)$, after dividing all coordinates by $\max|x_i|$, we can choose the x_i so that $\max|x_i| = 1$. Now define

$$^\sim : \ \mathbb{P}^n(k) \longrightarrow \mathbb{P}^n(\mathfrak{o}/\mathfrak{p}), \quad \mathbf{x} \longmapsto (\tilde{x}_i). \qquad (7.3.2)$$

We shall assume that the coefficients of $F(X)$ are in $\mathfrak{o}$. This can always be made to be true by a linear change in X. Then $\tilde{C}$ is defined to be the curve over $\mathfrak{o}/\mathfrak{p}$ obtained by replacing each coefficient f_i by $\tilde{f}_i$. Similarly for the reduction of the jacobian $\tilde{J}$. Then reduction gives a map from $C(k)$ to $\tilde{C}(\mathfrak{o}/\mathfrak{p})$ and from $\mathfrak{G}$ to $\tilde{\mathfrak{G}}$, where

$$\mathfrak{G} = J(k), \quad \tilde{\mathfrak{G}} = \tilde{J}(\mathfrak{o}/\mathfrak{p}). \qquad (7.3.3)$$

Note that our definition of the reduction map is dependent on the choice of model. There is a discussion of genus 2 minimal models in Liu (1994), but this will not be required here since our intended applications in Sections 4, 5 will be on finite extensions of $\mathbb{Q}_p$, and we shall make do entirely with model-dependent definitions. If the discriminant of $F(X)$ is a unit in $\mathfrak{o}$ then this will be a surjective group homomorphism, otherwise we shall regard it merely as a map between varieties. In either case, we can define the *kernel of reduction*, $\mathfrak{G}^0$, by¶

$$\mathfrak{G}^0 = \{z \in \mathfrak{G} : \ \tilde{z} = \tilde{\mathfrak{O}}\} \leqslant \mathfrak{G}. \qquad (7.3.4)$$

¶ Here and elsewhere we shall feel free to use the notation $H \leqslant G$ (standard in group theory, but often avoided in number theory texts) to mean that H is a subgroup of G.

Clearly, as a set, $\mathfrak{G}^0$ is the same as the neighbourhood $\mathcal{N}$ in (1.16). Let $\mathcal{F}$ be the formal group induced by J, as described in Section 1. From Theorem 1.1, $\mathcal{F}$ makes the set of parameters in $\mathfrak{p}$ into a group, denoted $\mathcal{F}(\mathfrak{p})$. Then the isomorphism

$$\mathfrak{G}^0 \cong \mathcal{F}(\mathfrak{p}) \tag{7.3.5}$$

is immediate when J is either $J(\mathcal{E})$ or $J(\mathcal{C})$.

4   Torsion in the kernel of reduction.

For this section, we shall assume that the ground field k is a finite extension of some $\mathbb{Q}_p$, where $p \neq \infty$. It follows that the maximal ideal $\mathfrak{p}$ must be the prime ideal given by $\pi\mathfrak{o}$, for some $\pi \in \mathfrak{p}$, and that the residue field $\mathfrak{o}/\mathfrak{p}$ is finite. We shall also assume that all coefficients of $F(X)$ are in $\mathfrak{o}$. The isomorphism between $\mathfrak{G}^0$ and $\mathcal{F}(\mathfrak{p})$ restricts what torsion can occur in $\mathfrak{G}^0$. For any m coprime to p the map $[m]$ on $\mathcal{F}$ has linear term ms. But m is invertible in $\mathfrak{o}$ and so an induction on the degree shows the existence of a power series over $\mathfrak{o}$ which is the inverse of $[m]$. Therefore, $[m]$ is an isomorphism from $\mathcal{F}(\mathfrak{p})$ to itself, so that $\mathcal{F}(\mathfrak{p})$ does not have any m torsion. It remains to determine what p power torsion can occur. Suppose that $\mathbf{s} \in \mathcal{F}(\mathfrak{p})$ is torsion of order p^n. As usual, the one-parameter argument in [Silverman (1986), p. 124] carries over directly to deduce from the corollary to Lemma 2.1, using induction on n, that

$$v(\mathbf{s}) \leqslant \frac{1}{p^n - p^{n-1}}, \tag{7.4.1}$$

where $v(\mathbf{s}) = \max(v(s_i))$, $v(s_i) = \log(|s_i|)$, normalized so that $v(p) = 1$.

THEOREM 7.4.1. *Let J be defined over $\mathfrak{o}$, the valuation ring of a finite extension of $\mathbb{Q}_p$, where $p \neq \infty$. The only torsion in $\mathfrak{G}^0$, the kernel of reduction, is p power torsion. If $\mathfrak{A} \in \mathfrak{G}^0$ is a torsion element of order p^n then the local parameter $\mathbf{s}$ corresponding to $\mathfrak{A}$ satisfies (1).*

Corollary. *Let $k = \mathbb{Q}_p$, $\mathfrak{o} = \mathbb{Z}_p$, $p \neq \infty$. Then $\mathfrak{G}^0$ is torsion free.*

5   The order of $\mathfrak{G}/2\mathfrak{G}$ when k is a finite extension of $\mathbb{Q}_p$, $p \neq \infty$.

The same conditions on k, $\mathfrak{o}$ and $\mathfrak{p} = \pi\mathfrak{o}$ apply as stated at the beginning of Section 4, and we continue to assume that the coefficients of $F(X)$ lie in $\mathfrak{o}$. Recall that $\mathfrak{G}^0$, the kernel of reduction, is a subgroup of $\mathfrak{G}$. We first wish to show that $\mathfrak{G}^0$ is of finite index in $\mathfrak{G}$. In the case where the discriminant of $F(X)$ is a unit in $\mathfrak{o}$ this is immediate, since the index of $\mathfrak{G}^0$ in $\mathfrak{G}$ is just the order of the group $\widetilde{\mathfrak{G}}$. With no restriction on the discriminant of $F(X)$, we can argue as follows. First note that k is locally compact, and so $\mathbb{P}^n(k)$ is compact. The set $\mathfrak{G}$, in either the dimension 1 or 2 situation, is a closed subset of $\mathbb{P}^n(k)$ and so is compact also. The set $\mathfrak{G}^0$, which is the same as the set $\mathcal{N}$ of (1.16), is clearly open. Furthermore, any coset $\mathfrak{A} + \mathfrak{G}^0$ is open, since addition is continuous. If $\mathfrak{G}^0$ were of infinite index in $\mathfrak{G}$ then there would be the contradiction that the cosets of $\mathfrak{G}^0$ would give a disjoint open covering of $\mathfrak{G}$. It follows that

$$[\mathfrak{G} : \mathfrak{G}^0] < \infty, \qquad (7.5.1)$$

regardless of whether the discriminant of $F(X)$ is a unit. The group $\mathfrak{G}^0$ is isomorphic to $\mathcal{F}(\mathfrak{p})$, and the identity map

$$\mathcal{F}(\mathfrak{p}^n)/\mathcal{F}(\mathfrak{p}^{n+1}) \longrightarrow \mathfrak{p}^n/\mathfrak{p}^{n+1} \qquad (7.5.2)$$

is an isomorphism to a finite group. It follows by induction that the group $\mathcal{F}(\mathfrak{p}^n)$ is of finite index in $\mathcal{F}(\mathfrak{p})$ for any n. In particular, since k is a finite extension of $\mathbb{Q}_p$, the rational prime $p = \pi^n$, for some n, so that

$$[\mathcal{F}(\mathfrak{p}) : \mathcal{F}(p \cdot \mathfrak{o})] < \infty. \qquad (7.5.3)$$

The estimate

$$|m!| > p^{-m/(p-1)}, \qquad (7.5.4)$$

together with (2.6), implies that the formal logarithm and exponential functions both converge on $p \cdot \mathfrak{o}$, giving an isomorphism

$$L : \quad \mathcal{F}(p \cdot \mathfrak{o}) \longrightarrow \mathcal{F}^+(p \cdot \mathfrak{o}), \qquad (7.5.5)$$

where $\mathcal{F}^+$ is the additive formal group law $\mathbf{s} + \mathbf{t}$. But the group $\mathcal{F}^+(p \cdot \mathfrak{o})$ is isomorphic to the additive group $p \cdot \mathfrak{o}$ in the elliptic curve case of $J = J(\mathcal{E})$, and to the additive group $p \cdot \mathfrak{o} \times p \cdot \mathfrak{o}$ when $J = J(\mathcal{C})$ and $\mathcal{C}$ is of genus 2. Since the group $p \cdot \mathfrak{o}$ is isomorphic to $\mathfrak{p}$, we have shown the following special case of Mattuck (1955).

THEOREM 7.5.1. *Let J be defined over k, a finite extension of $\mathbb{Q}_p$, where $p \neq \infty$. Then $\mathfrak{G}$ contains a subgroup $\mathfrak{H}$ of finite index such that $\mathfrak{H}$ is isomorphic to the additive group $\mathfrak{p}$ when $J = J(\mathcal{E})$, and $\mathfrak{p} \times \mathfrak{p}$ when $J = J(\mathcal{C})$.*

The existence of such a subgroup gives a way of expressing the order of $\mathfrak{G}/2\mathfrak{G}$ in terms of the order of $\mathfrak{G}[2]$, the 2-torsion subgroup of $\mathfrak{G}$. First consider the multiplication by 2 map from the finite group $\mathfrak{G}/\mathfrak{H}$

onto $2\mathfrak{G}/2\mathfrak{H}$. This has kernel $\mathfrak{G}[2] + \mathfrak{H}$, which has the same order as $\mathfrak{G}[2]$, since $\mathfrak{H}$ is torsion free. This gives¶

$$\#\mathfrak{G}/\mathfrak{H} = \#2\mathfrak{G}/2\mathfrak{H} \cdot \#\mathfrak{G}[2]. \tag{7.5.6}$$

The natural surjections from $\mathfrak{G}/2\mathfrak{H}$ onto $\mathfrak{G}/\mathfrak{H}$ with kernel $\mathfrak{H}/2\mathfrak{H}$ and from $\mathfrak{G}/2\mathfrak{H}$ onto $\mathfrak{G}/2\mathfrak{G}$ with kernel $2\mathfrak{G}/2\mathfrak{H}$ give the identities

$$\#\mathfrak{G}/2\mathfrak{H} = \#\mathfrak{G}/\mathfrak{H} \cdot \#\mathfrak{H}/2\mathfrak{H}, \quad \#\mathfrak{G}/2\mathfrak{H} = \#\mathfrak{G}/2\mathfrak{G} \cdot \#2\mathfrak{G}/2\mathfrak{H}, \tag{7.5.7}$$

respectively. Combining these equations, together with the fact that $\mathfrak{H}$ is isomorphic to $\mathfrak{p}$, gives the key result

$$\#\mathfrak{G}/2\mathfrak{G} = \#\mathfrak{G}[2] \cdot \#\mathfrak{p}/2\mathfrak{p}. \tag{7.5.8}$$

This last identity makes it clear why extensions of $\mathbb{Q}_p$ will behave differently in the cases $p = 2$ and $p \neq 2$. In the former case, $2\mathfrak{p}$ will be a proper subgroup of $\mathfrak{p}$, whereas in the latter case 2 is a unit and so $\mathfrak{p} = 2\mathfrak{p}$. Some readers may prefer to replace equations (6), (7), (8) with the following Snake Lemma diagram:

$$
\begin{array}{ccccccccc}
 & & & & & & 0 & & \\
 & & & & & & \downarrow & & \\
 & & 0 & \longrightarrow & \mathfrak{G}[2] & \longrightarrow & \mathrm{Ker} & & \\
 & & \downarrow & & \downarrow & & \downarrow & & \\
0 & \longrightarrow & \mathfrak{H} & \longrightarrow & \mathfrak{G} & \longrightarrow & \mathfrak{G}/\mathfrak{H} & \longrightarrow & 0 \\
 & & \downarrow{\scriptstyle[2]} & & \downarrow{\scriptstyle[2]} & & \downarrow{\scriptstyle[2]} & & \\
0 & \longrightarrow & \mathfrak{H} & \longrightarrow & \mathfrak{G} & \longrightarrow & \mathfrak{G}/\mathfrak{H} & \longrightarrow & 0 \\
 & & \downarrow & & \downarrow & & \downarrow & & \\
 & & \mathfrak{H}/2\mathfrak{H} & \longrightarrow & \mathfrak{G}/2\mathfrak{G} & \longrightarrow & \mathrm{Coker} & \longrightarrow & 0 \\
 & & & & & & \downarrow & & \\
 & & & & & & 0 & &
\end{array}
\tag{7.5.9}
$$

¶ To avoid later becoming overwhelmed with brackets we immediately get the reader used to the style $\#\mathfrak{G}/\mathfrak{H}$ to mean $\#(\mathfrak{G}/\mathfrak{H})$, the order of the group $\mathfrak{G}/\mathfrak{H}$; similarly (cf. Chapter 10, Section 4) we shall use the expression $\#\widehat{\mathfrak{G}}/\phi(\mathfrak{G}) \cdot \#\mathfrak{G}/\hat{\phi}(\widehat{\mathfrak{G}})$ to mean $\#(\widehat{\mathfrak{G}}/\phi(\mathfrak{G})) \cdot \#(\mathfrak{G}/\hat{\phi}(\widehat{\mathfrak{G}}))$, and so on. Note that, in all cases, the intended interpretation the only one which makes sense.

where the entries Ker and Coker should be defined in the unique way which makes their column exact; that is, they are defined to be

$$\ker\left(\mathfrak{G}/\mathfrak{H}\xrightarrow{[2]}\mathfrak{G}/\mathfrak{H}\right) \quad \text{and} \quad \left(\mathfrak{G}/\mathfrak{H}\right)/\mathrm{im}\left(\mathfrak{G}/\mathfrak{H}\xrightarrow{[2]}\mathfrak{G}/\mathfrak{H}\right), \qquad (7.5.10)$$

respectively. Since $\mathfrak{G}/\mathfrak{H}$ is finite we have #Ker = #Coker. The Snake Lemma allows us to join the top and bottom nontrivial rows into a single exact sequence, in which we can equate the product of the orders of Ker, $\mathfrak{G}/2\mathfrak{G}$ with the product of the orders of $\mathfrak{G}[2]$, $\mathfrak{H}/2\mathfrak{H}$, Coker, and so (8) again follows.

We may deduce the following in the special case of $k = \mathbb{Q}_p$.

Corollary. *Let J be defined over $\mathbb{Q}_p$, where $p \neq \infty$. Let $\mathfrak{G}[2]$ denote the 2-torsion subgroup of $\mathfrak{G}$. If $p \neq 2$ then $\#\mathfrak{G}/2\mathfrak{G} = \#\mathfrak{G}[2]$. Otherwise, $\#\mathfrak{G}/2\mathfrak{G} = \#\mathfrak{G}[2] \cdot 2^g$, where $g = 1$ when $J = J(\mathcal{E})$ and $g = 2$ when $J = J(\mathcal{C})$.*

The above discussion can be generalized to abelian varieties of dimension d for which the above identities remain true, where $\mathfrak{H}$ is isomorphic to d copies of $\mathfrak{p}$.

6   The order of $\mathfrak{G}/2\mathfrak{G}$ when $k = \mathbb{R}$ or $\mathbb{C}$.

When the sextic $F(X)$, which occurs in $\mathcal{C} : Y^2 = F(X)$, is defined over $\mathbb{R}$ there must at least be a factorization

$$F(X) = q_1(X)q_2(X)q_3(X), \qquad (7.6.1)$$

where each $q_i(X)$ is defined over $\mathbb{R}$. Therefore, when $\mathfrak{G} = J(\mathbb{R})$, the 2-torsion group $\mathfrak{G}[2]$ contains at least the 4 elements $\mathfrak{O}$ and $\{(a_i, 0), (a_i', 0)\}$, $i = 1, 2, 3$, where a_i and a_i' are the roots of $q_i(X)$.

In the notation of Chapter 6, the map $\mu : \mathfrak{G} \to \mathcal{M}$ has kernel generated by $2\mathfrak{G}$ and $\mathfrak{W}$. If at least one $q_i(X)$ has a root in $\mathbb{R}$ then Criterion (i) of Lemma 6.5.1 is satisfied; otherwise Criterion (ii) is satisfied by the set $\{(a_1, 0), (a_2, 0), (a_3, 0)\}$. In either case we can conclude that $\mathfrak{W} \subset 2\mathfrak{G}$. Therefore the induced map $\bar{\mu}$ is injective on $\mathfrak{G}/2\mathfrak{G}$. If at most one of the $q_i(X)$ splits in $\mathbb{R}$ then the image of μ is the trivial group, as must be $\mathfrak{G}/2\mathfrak{G}$. Note that in this case $\mathfrak{G}[2]$ has order exactly 4. If two or more of the $q_i(X)$ split then the resulting real roots break $\mathbb{R}$ into a finite number of intervals. The image of $\{(x, y), (u, v)\}$ under μ, in the case when both points are real, is determined by which of these intervals contain each of (x, y) and (u, v). Note that the image under μ is always 1 if the points are complex and conjugate. This gives a finite number of cases to check to show that the image of μ, and hence $\mathfrak{G}/2\mathfrak{G}$, has order 2 when exactly two of the quadratics split, and has order 4 when all three of the quadratics split. In all possible cases it follows that $\mathfrak{G}/2\mathfrak{G}$ is generated by two fewer elements than $\mathfrak{G}[2]$. An almost identical argument for the elliptic curve situation shows that $\mathfrak{G}/2\mathfrak{G}$ is generated by one less element than $\mathfrak{G}[2]$. This can all be summarized by

$$\#\mathfrak{G}/2\mathfrak{G} = \#\mathfrak{G}[2]/2^g, \tag{7.6.2}$$

where $g = 1$ when $J = J(\mathcal{E})$ and $g = 2$ when $J = J(\mathcal{C})$.

The identities (2) and (5.8) give a method for computing $\mathfrak{G}/2\mathfrak{G}$ when $\mathfrak{G} = J(k)$, for any k which is the completion of a number field. The problem is reduced to that of finding the irreducible factorization of $F(X)$ over k. From this, one can easily determine the order of $\mathfrak{G}[2]$ and hence the order of $\mathfrak{G}/2\mathfrak{G}$. The only case which we have not so far mentioned explicitly is $k = \mathbb{C}$; but this is trivial since the map [2] is surjective, and so $\#\mathfrak{G}/2\mathfrak{G} = 1$.

Chapter 8

Torsion

0   Introduction. We shall consider two computational problems relating to the torsion group of the jacobian. First, given a curve of genus 2 defined over $\mathbb{Q}$, how does one try to find the rational torsion group on the jacobian? Second, given an integer N, how does one try to find a curve whose jacobian has a rational point of order N? Neither of these questions has been adequately answered, but we give an idea of the general strategy. For the first question, we shall give a method which works well in practice, and there is a crude algorithm which should be amenable to improvement. Most of the recent work on the second question comes down to a bag of tricks for searching for large torsion orders; we shall give the one which has proved successful in finding most torsion orders up to 29 over $\mathbb{Q}$. There is not yet any method for bounding the possible torsion which can occur.

1   Computing the group law. Since the next few chapters are to be devoted largely to computational techniques, we shall first illustrate the mechanical details involved in computing the group law on divisors in more detail than the description in Chapter 1. Any reader who is familiar with addition of divisors modulo linear equivalence can safely skip this section. We shall not be requiring any of the structure of the jacobian as a variety here, but are only interested in the group structure. The points on $J = J(\mathcal{C})$ may be identified with those on Pic^2, as introduced in Chapter 1, Section 1. Consider, for example, the following curve defined over $\mathbb{Q}$.

$$\mathcal{C}_1: \quad Y^2 = F_1(X) = X(X-1)(X-2)(X-5)(X-6). \qquad (8.1.1)$$

The *Weierstrass points* are the six points which are invariant under the *hyperelliptic involution* $(x, y) \mapsto (x, -y)$. In this case, they are: $(0,0), (1,0),$ $(2,0), (5,0), (6,0)$ and ∞, the point at infinity. We shall use $\{(x,y), (u,v)\}$ as a shorthand notation for the divisor class containing $(x,y) + (u,v) - 2 \cdot \infty$; it is allowable for $(x,y) = (u,v)$. This gives a way of describing points on the jacobian, except that all pairs of the form $\{(x,y), (x,-y)\}$, together with $\{\infty, \infty\}$, must be identified into the single element $\mathfrak{O}$. Note that, for example, $\{(1,0),(1,0)\} = \mathfrak{O}$. With this representation, the members of the Mordell-Weil group $\mathfrak{G} = J(\mathbb{Q})$ must satisfy

$$(x,y),(u,v) \in C_1(\mathbb{Q})$$

$$\text{or} \tag{8.1.2}$$

$$(x,y),(u,v) \in C_1(\mathbb{Q}(\sqrt{d})) \text{ with } (x,y) \text{ and } (u,v) \text{ conjugate over } \mathbb{Q},$$

where $d \in \mathbb{Q}^*$ and $d \notin (\mathbb{Q}^*)^2$. For example, $\{\infty,(1,0)\}$, $\{(0,0),(10,-120)\}$, $\{(\frac{11}{2} + \frac{1}{2}\sqrt{41}, 35 + 5\sqrt{41}), (\frac{11}{2} - \frac{1}{2}\sqrt{41}, 35 - 5\sqrt{41})\}$, $\{(3,6),(3,6)\}$ are all in $\mathfrak{G}$. But $\{(3,6),(\frac{11}{2} + \frac{1}{2}\sqrt{41}, 35 + 5\sqrt{41})\}$ is not in $\mathfrak{G}$ since the two points are not conjugate over $\mathbb{Q}$. Consider $\{(0,0),(1,0)\} \in \mathfrak{G}$. Then

$$\{(0,0),(1,0)\} + \{(0,0),(1,0)\} = \{(0,0),(0,0)\} + \{(1,0),(1,0)\}$$
$$= \mathfrak{O} + \mathfrak{O} = \mathfrak{O}, \tag{8.1.3}$$

so that $\{(0,0),(1,0)\}$ is a torsion divisor of order 2. Clearly the 15 members of $\mathfrak{G}$ which are of order 2 are given by $\{P_1,P_2\}$, where P_1, P_2 are distinct Weierstrass points. Together with $\mathfrak{O}$ these give the full 2-torsion subgroup of $\mathfrak{G}$.

For a more typical addition, let $\mathfrak{A} = \{(0,0),(1,0)\}$, $\mathfrak{B} = \{(2,0),(3,6)\}$. Suppose that we wish to add $\mathfrak{A} + \mathfrak{B}$. The first step is to find the unique function $Y = $ (cubic in X) which meets C_1 at these four points, namely

$$Y = X(X-1)(X-2). \tag{8.1.4}$$

Substituting this for Y in $Y^2 - F_1(X)$ gives a polynomial in X which factors as

$$X(X-1)(X-2)(X-3)(X^2 - X + 10). \tag{8.1.5}$$

The roots $X = 0,1,2,3$ correspond to the X-coordinates of the points in the sets representing $\mathfrak{A}$ and $\mathfrak{B}$. The roots of the remaining quadratic can be substituted into the cubic of (4) to give the corresponding Y-coordinates. Therefore the divisor of the function (4) gives on $\mathfrak{G}$ that

$$\mathfrak{A} + \mathfrak{B} + \{(\frac{1}{2} + \frac{1}{2}\sqrt{-39}, 15 - 5\sqrt{-39}), (\frac{1}{2} - \frac{1}{2}\sqrt{-39}, 15 + 5\sqrt{-39})\}$$
$$= \mathfrak{O}, \tag{8.1.6}$$

and so, on negating Y-coordinates,

$$\mathfrak{A} + \mathfrak{B} =$$
$$\{(\frac{1}{2} + \frac{1}{2}\sqrt{-39}, -15 + 5\sqrt{-39}), (\frac{1}{2} - \frac{1}{2}\sqrt{-39}, -15 - 5\sqrt{-39})\}. \tag{8.1.7}$$

In this context (when the genus 2 curve is in quintic form) the presence of ∞ in a set representing one of the addends forces the coefficient of X^3 in the cubic to be 0. A point repeated m times, as with elliptic curves, forces the function $Y = $ (cubic in X) to meet C_1 with multiplicity m. For example, let $\mathfrak{C} = \{\infty, (0,0)\}$ and $\mathfrak{D} = \{(3,6), (3,6)\}$. Then our function $Y = $ (cubic in X) must have 0 as the coefficient of X^3, must pass through $(0,0)$ and $(3,6)$, and have the same derivative as C_1 at $(3,6)$, namely $y' = 3$. These conditions are uniquely satisfied by

$$Y = \frac{1}{3}X^2 + X. \tag{8.1.8}$$

Substituting this into $Y^2 - F_1(X)$, and repeating the previous process then gives

$$\mathfrak{C} + \mathfrak{D} = \{P, P'\}, \quad P = (\frac{73}{18} + \frac{1}{18}\sqrt{3169}, -\frac{3110}{243} - \frac{50}{243}\sqrt{3169}), \tag{8.1.9}$$

where P' is the conjugate of P over $\mathbb{Q}$. This gives an idea of how quickly the size of the integers can increase as repeated additions are performed.

A typical curve of genus 2 in the form $Y^2 = $ (sextic in X) can be birationally mapped over the ground field to the form $Y^2 = $ (quintic in X) if the sextic has a rational root; that is to say, if there is a rational Weierstrass point. This can be seen by mapping

$$Y^2 = (X - e_1)H(X), \tag{8.1.10}$$

where $H(X)$ is a quintic, to the curve

$$V^2 = U^5 H(e_1 + \frac{1}{U}) \tag{8.1.11}$$

using the birational map $U = 1/(X - e_1)$, $V = Y/(X - e_1)^3$. When the sextic does not have a rational root then no such map is possible. Consider, for example, the curve

$$C_2 : \quad Y^2 = F_2(X) = (X^2 + 1)(X^2 + 2)(X^2 + X + 1). \tag{8.1.12}$$

None of the six Weierstrass points (corresponding to the six roots of the sextic) are rational, and so the curve must remain in sextic form. We use the notation ∞^+ and ∞^- to represent the two branches at infinity which, for computational purposes, should be viewed as separate members of $C_2(\mathbb{C})$. Since the coefficient of X^6 is a square, these should both be regarded as points in $C_2(\mathbb{Q})$. In this context, the notation $\{(x,y), (u,v)\}$ is shorthand for the divisor class containing $(x, y) + (u, v) - \infty^+ - \infty^-$. The conditions for membership in $\mathfrak{G} = J(\mathbb{Q})$ are the same as before. For example, $\{\infty^+, \infty^+\} \in \mathfrak{G}$. There are three members of $\mathfrak{G}$ of torsion order 2, namely $\{(i, 0), (-i, 0)\}$, $\{(\sqrt{-2}, 0)(-\sqrt{-2}, 0)\}$ and $\{(-\frac{1}{2} + \frac{\sqrt{-3}}{2}, 0), (-\frac{1}{2} - \frac{\sqrt{-3}}{2}, 0)\}$, corresponding to the $\mathbb{Q}$-rational quadratic factors of $F_2(X)$. Together with $\mathfrak{O}$, this gives a 2-torsion group of order 4 in $\mathfrak{G}$. The other 2-torsion divisors, such as $\{(i, 0), (\sqrt{-2}, 0)\}$ are not in $\mathfrak{G}$. The 'new' type of addition here appears when one of the branches at infinity occurs. For example, suppose we want to add $\mathfrak{E} + \mathfrak{F}$, where $\mathfrak{E} = \{(i, 0), (-i, 0)\}$ and $\mathfrak{F} = \{\infty^+, \infty^+\}$. Then the presence of ∞^+ forces the coefficient of X^3 in our cubic to be 1 (this would have been -1 if ∞^+ had been replaced by ∞^-). The second occurrence of ∞^+ means that we must choose the coefficient of X^2 so that our (cubic in $X)^2 - F_2(X)$ is only of degree 4; this forces that coefficient to be $\frac{1}{2}$. The points $(i, 0)$ and $(-i, 0)$ determine the remaining two coefficients, giving the function

$$Y = (X + \frac{1}{2})(X^2 + 1). \tag{8.1.13}$$

Proceeding as usual gives

$$\mathfrak{E} + \mathfrak{O} = \{Q, Q'\}, \quad Q = (-\frac{2}{7} + \frac{3}{7}\sqrt{-5}, \frac{192}{343} + \frac{6}{343}\sqrt{-5}), \tag{8.1.14}$$

where Q' is the conjugate of Q over $\mathbb{Q}$.

2   Computing the torsion of a given jacobian. A standard technique, as with elliptic curves, is to use the reduction map modulo a prime not dividing the discriminant. Suppose that $\mathcal{C} : Y^2 = F(X)$ is a curve of genus 2, where we shall assume that the ground field is $\mathbb{Q}$ (so that $\mathfrak{G} = J(\mathbb{Q})$), although it is straightforward to generalize the technique to number fields. After adjusting, if necessary, X and Y by a constant, we can assume that $F(X)$ is defined over $\mathbb{Z}$. Let p be a prime not dividing $2\,\mathrm{disc}\,(F)$, where $\mathrm{disc}\,(F)$ is the discriminant of $F(X)$. Then $\tilde{\mathcal{C}}$, the reduction of $\mathcal{C}$ modulo p, is a curve of genus 2 over $\mathbb{F}_p$, with jacobian $\tilde{J}$. Recall from the corollary to Theorem 7.4.1 that the reduction map

$$\sim \; : \quad \mathfrak{G} \longrightarrow \tilde{J}(\mathbb{F}_p) \tag{8.2.1}$$

is injective on the torsion group $\mathfrak{G}_{\mathrm{tors}}$. For each choice of p, finding the group $\tilde{J}(\mathbb{F}_p)$ is only a finite computation, and it must contain a copy of $\mathfrak{G}_{\mathrm{tors}}$ as a subgroup. The following is then immediate.

Lemma 8.2.1. *The group* $\mathfrak{G} = J(\mathbb{Q})$ *has finite torsion subgroup* $\mathfrak{G}_{\text{tors}}$.

One hopes that after a few primes, the resulting bound on the order of $\mathfrak{G}_{\text{tors}}$ is the same as the number of known torsion points.

Example 8.2.1. *Let* $C_1 : Y^2 = F_1(X)$ *be the curve of (1.1). Then* $\mathfrak{G}_{\text{tors}}$ *is of order 16 and consists entirely of 2-torsion.*

Proof. The discriminant $\text{disc}(F_1)$ is $2^{12} 3^4 5^4$ so that $\widetilde{C}_1$ is a curve of genus 2 over $\mathbb{F}_p$ for any prime $p \neq 2, 3, 5$. Take $p = 7$. The set $\widetilde{C}_1(\mathbb{F}_7)$ consists of the six Weierstrass points (including ∞) together with $(3, 6), (3, -6)$. The resulting members of $\widetilde{J}(\mathbb{F}_7)$ are $\mathfrak{O}$, $\{P_1, P_2\}$, where P_1, P_2 are distinct Weierstrass points (giving all 15 points of order 2), $\pm\{P, (3, 6)\}$, where P is a Weierstrass point, and $\pm\{(3, 6), (3, 6)\}$. These give $1 + 15 + 12 + 2 = 30$ members of $\widetilde{J}(\mathbb{F}_7)$. Note that we do not count, for example, such sets as $\{(0, 0), (0, 0)\}$ or $\{(3, 6), (3, -6)\}$ since these are in the class $\mathfrak{O}$. We now search for all points in $\widetilde{C}_1(\mathbb{F}_{49})$. Writing members of $\mathbb{F}_{49}$ as $a + b\sqrt{3}$, where $a, b \in \mathbb{F}_7$, a finite search finds 36 points of the form $(a + b\sqrt{3}, c + d\sqrt{3})$, with $b \neq 0$. When these are put into conjugate pairs, they contribute nine members of $\widetilde{J}(\mathbb{F}_7)$ of the form $\{(a + b\sqrt{3}, c + d\sqrt{3}), (a - b\sqrt{3}, c - d\sqrt{3})\}$. Note that we can discount $b = 0$ since any such would contribute either a member of $\widetilde{C}(\mathbb{F}_7)$ (already counted), or the member $\{(a, d\sqrt{3}), (a, -d\sqrt{3})\}$ of the class $\mathfrak{O}$ (which has also already been counted). So, in total we have

$$\#\widetilde{J}(\mathbb{F}_7) = 48. \tag{8.2.2}$$

Repeating the above process for $p = 11$ gives

$$\#\widetilde{J}(\mathbb{F}_{11}) = 176. \tag{8.2.3}$$

But from the injection of (1) it follows that $\#\mathfrak{G}_{\text{tors}}$ divides both 48 and 176, and so divides 16, the greatest common divisor, as required.

Now let C be a general curve $Y^2 = F(X)$ defined over $\mathbb{Z}$, with p not dividing $2\,\text{disc}(F)$, and fix γ, a quadratic non-residue modulo p. If

$$w_p = \#W_p, \; r_p = \#R_p, \; t_p = \#T_p,$$

where

$$W_p = \{P \in \widetilde{C}(\mathbb{F}_p) : \; P \text{ is a Weierstrass point}\},$$
$$R_p = \{P \in \widetilde{C}(\mathbb{F}_p) : \; P \text{ not a Weierstrass point}\}, \tag{8.2.4}$$
$$T_p = \{(a + b\sqrt{\gamma}, c + d\sqrt{\gamma}) \in \widetilde{C}(\mathbb{F}_{p^2}) : \; b \neq 0\},$$

then $\widetilde{J}(\mathbb{F}_p)$ will have $\frac{1}{2} w_p(w_p - 1)$ members of the form $\{P_1, P_2\}$, $\quad w_p r_p$ members of the form $\{P_1, Q_1\}$, and $\frac{1}{2} r_p^2$ members of the form $\{Q_1, Q_2\}$, where $P_1, P_2 \in W_p$ and $Q_1, Q_2 \in R_p$. Adding in 1 for $\mathfrak{O}$ and $\frac{1}{2}t_p$ for the conjugate pairs contributed by T_p gives

$$\#\widetilde{J}(\mathbb{F}_p) = 1 + \frac{1}{2}w_p(w_p - 1) + w_p r_p + \frac{1}{2}r_p^2 + \frac{1}{2}t_p. \qquad (8.2.5)$$

Note that, in the above example, the point at infinity ∞ was included in the set W_p, since the curve was in quintic form. When the curve is in sextic form, and the coefficient of X^6 is a quadratic residue, then ∞^+, ∞^- give two distinct members of R_p. When the coefficient of X^6 is a non-residue then ∞^+, ∞^- should be viewed as irrational, and do not make any contribution.

There is a slightly more elegant expression for $\#\widetilde{J}(\mathbb{F}_p)$ which can be derived from the observation that

$$\#\widetilde{C}(\mathbb{F}_p) = r_p + w_p, \quad \#\widetilde{C}(\mathbb{F}_{p^2}) = t_p + 2(p+1) - w_p. \qquad (8.2.6)$$

Substituting these into (5) gives¶

$$\#\widetilde{J}(\mathbb{F}_p) = \frac{1}{2}\#\widetilde{C}(\mathbb{F}_{p^2}) + \frac{1}{2}\left(\#\widetilde{C}(\mathbb{F}_p)\right)^2 - p, \qquad (8.2.7)$$

where, of course, in $\#\widetilde{C}(\mathbb{F}_p)$, we must count the point at infinity once if the coefficient of X^6 is 0 mod p, twice if the coefficient is a nonzero quadratic residue, and otherwise no times. In $\#\widetilde{C}(\mathbb{F}_{p^2})$, we again must count the point at infinity once if the coefficient of X^6 is 0 mod p, and otherwise twice, since the coefficient is guaranteed to be a quadratic residue in $\mathbb{F}_{p^2}$.

In practice, it nearly always seems to happen that $\mathfrak{G}_{\text{tors}}$ can be found, as in the example above, by taking the greatest common divisor of sufficiently many $\#\widetilde{J}(\mathbb{F}_p)$ such that p does not divide $2\,\mathrm{disc}\,(F)$. However, there are occasions when this is not sufficient and it is necessary to use the fact that the map given in (1) is a homomorphism of groups.

Example 8.2.2. *Let $C_2 : Y^2 = F_2(X)$ be the curve of (1.12). Then $\mathfrak{G}_{\text{tors}}$ consists of the 2-torsion group of order 4, even though $8 | \#\widetilde{J}(\mathbb{F}_p)$ for all p not dividing $2\,\mathrm{disc}\,(F)$.*

Proof. The discriminant $\mathrm{disc}\,(F_2)$ is $-2^5 3^3$, so that $\widetilde{C}_2$ is a curve of genus 2 over $\mathbb{F}_p$ for any prime $p \neq 2, 3$. The 2-torsion subgroup of $\mathfrak{G}_{\text{tors}}$ is of order 4; namely, $\mathfrak{O}$ and the elements $\mathfrak{A}_1, \mathfrak{A}_2, \mathfrak{A}_3$ corresponding to the $\mathbb{Q}$-rational

---

¶ This formula can also be obtained by evaluating at $X = 1$ the characteristic polynomial of the Frobenius endomorphism given in Lemma 3 of Merriman and Smart (1993), namely $X^4 - tX^3 + sX^2 - ptX + p^2$, where $t = p + 1 - \#\widetilde{C}(\mathbb{F}_p)$ and $s = \frac{1}{2}(\#\widetilde{C}(\mathbb{F}_p)^2 + \#\widetilde{C}(\mathbb{F}_{p^2})) + p - (p+1)\#\widetilde{C}(\mathbb{F}_p)$.

quadratic factors $X^2 + 1$, $X^2 + 2$, $X^2 + X + 1$, respectively, of $F_2(X)$. For $p = 5$, we have $\#\widetilde{C}(\mathbb{F}_5) = 6$ and $\#\widetilde{C}(\mathbb{F}_{25}) = 22$; we can use (7) to deduce that $\#\widetilde{J}(\mathbb{F}_5) = 24$. It is a finite amount of work to check further that $\widetilde{J}(\mathbb{F}_5) \cong C_2 \times C_3 \times C_4$ and that $\widetilde{\mathfrak{A}}_1 \in 2\widetilde{J}(\mathbb{F}_5)$. Similarly, $\#\widetilde{J}(\mathbb{F}_7) = 64$. At this point, we can conclude that

$$\#\mathfrak{G}_{\text{tors}} = 8 \text{ if } \widetilde{\mathfrak{A}}_1 \in 2\mathfrak{G}, \quad \#\mathfrak{G}_{\text{tors}} = 4 \text{ if } \widetilde{\mathfrak{A}}_1 \notin 2\mathfrak{G}. \tag{8.2.8}$$

This is unresolved by $p = 11$, where $\#\widetilde{J}(\mathbb{F}_{11}) = 80$ and $\widetilde{\mathfrak{A}}_1 \in 2\widetilde{J}(\mathbb{F}_{11})$. However, when $p = 13$, we have $\#\widetilde{J}(\mathbb{F}_{13}) = 176$ and $\widetilde{\mathfrak{A}}_1 \notin 2\widetilde{J}(\mathbb{F}_{13})$. This finally gives that $\widetilde{\mathfrak{A}}_1 \notin 2\mathfrak{G}$, using the fact that the map given in (1) is a homomorphism. Hence $\#\mathfrak{G}_{\text{tors}} = 4$. It only remains to show that $8 | \#\widetilde{J}(\mathbb{F}_p)$ for all p not dividing $2\,\text{disc}\,(F_2)$.

For a general prime p not dividing $2\,\text{disc}\,(F_2)$, suppose that $\widetilde{\mathfrak{A}}_1 \notin 2\widetilde{J}(\mathbb{F}_p)$. Consider the divisor

$$\mathfrak{B}_1 = \{(0, \gamma), (1, 3\gamma)\} \in J(\mathbb{F}_{p^2}), \tag{8.2.9}$$

where $\gamma^2 = 2$. One can verify directly that $2\mathfrak{B}_1 = \widetilde{\mathfrak{A}}_1$.

Let σ represent the nontrivial automorphism of $\mathbb{F}_{p^2}$ over $\mathbb{F}_p$. One of the following four cases must be satisfied (where residue means quadratic residue). We shall see that in each case, the group $\widetilde{J}(\mathbb{F}_p)$ contains a subgroup isomorphic to either $C_2 \times C_4$ or $C_2 \times C_2 \times C_2$.

Case 1. 2 *is a residue modulo p.* Here, $\sigma(\gamma) = \gamma$ and $\sigma(\mathfrak{B}_1) = \mathfrak{B}_1$. Therefore $\mathfrak{B}_1 \in \widetilde{J}(\mathbb{F}_p)$ and so $\widetilde{\mathfrak{A}}_1 \in 2\widetilde{J}(\mathbb{F}_p)$. The subgroup of $\widetilde{J}(\mathbb{F}_p)$ generated by $\mathfrak{B}_1$ and $\widetilde{\mathfrak{A}}_2$ is isomorphic to $C_2 \times C_4$.

Case 2. 2 *and -1 are both non-residues modulo p.* Fix a choice of $i \in \mathbb{F}_{p^2}$ such that $i^2 = -1$. Then neither i nor γ are in $\mathbb{F}_p$ and $\sigma(i) = -i$, $\sigma(\gamma) = -\gamma$. Consider the divisor

$$\mathfrak{B}_2 = \mathfrak{B}_1 + \{(i, 0), (i\gamma, 0)\}. \tag{8.2.10}$$

Then $\sigma(\mathfrak{B}_1) = -\mathfrak{B}_1 = \mathfrak{B}_1 - \widetilde{\mathfrak{A}}_1$ and $\sigma(\{(i, 0), (i\gamma, 0)\}) = \{(-i, 0), (i\gamma, 0)\} = \{(i, 0), (i\gamma, 0)\} + \widetilde{\mathfrak{A}}_1$. Hence, $\sigma(\mathfrak{B}_2) = \mathfrak{B}_2$ and so $\mathfrak{B}_2 \in \widetilde{J}(\mathbb{F}_p)$. Also $2\mathfrak{B}_2 = \widetilde{\mathfrak{A}}_1$ and so, as in case 1, $\widetilde{\mathfrak{A}}_1 \in 2\widetilde{J}(\mathbb{F}_p)$. The subgroup of $\widetilde{J}(\mathbb{F}_p)$ generated by $\mathfrak{B}_2$ and $\widetilde{\mathfrak{A}}_2$ is isomorphic to $C_2 \times C_4$.

Case 3. 2 *is a non-residue, -1 a residue, and -3 a non-residue modulo p.* Fix a choice of $\delta \in \mathbb{F}_{p^2}$ such that $\delta^2 = -3$. Then $i \in \mathbb{F}_p$, but $\gamma, \delta \notin \mathbb{F}_p$. Consider the divisor

$$\mathfrak{B}_3 = \mathfrak{B}_1 + \{(i\gamma, 0), (-\frac{1}{2} + \frac{1}{2}\delta, 0)\}. \tag{8.2.11}$$

By a similar argument to that in case 2, we see that $\sigma(\mathfrak{B}_3) = \mathfrak{B}_3$ and so $\mathfrak{B}_3 \in \tilde{J}(\mathbb{F}_p)$. Also $2\mathfrak{B}_3 = \tilde{\mathfrak{A}}_1$ and so we again have that $\mathfrak{A}_1 \in 2\tilde{J}(\mathbb{F}_p)$. The subgroup of $\tilde{J}(\mathbb{F}_p)$ generated by $\mathfrak{B}_3$ and $\tilde{\mathfrak{A}}_2$ is isomorphic to $C_2 \times C_4$.

Case 4. *2 is a non-residue, -1 and -3 are both residues modulo p.* This is the only case when $\tilde{\mathfrak{A}}_1 \notin \tilde{J}(\mathbb{F}_p)$. Since $\sigma(i) = i$ and $\sigma(\delta) = \delta$, the divisor

$$\mathfrak{B}_4 = \{(i,0), (-\frac{1}{2} + \frac{1}{2}\delta, 0)\} \tag{8.2.12}$$

gives a point of order 2 in $\tilde{J}(\mathbb{F}_p)$, independent of $\mathfrak{A}_1, \mathfrak{A}_2, \mathfrak{A}_3$. The subgroup of $\tilde{J}(\mathbb{F}_p)$ generated by $\mathfrak{B}_4, \tilde{\mathfrak{A}}_1, \tilde{\mathfrak{A}}_2$ is isomorphic to $C_2 \times C_2 \times C_2$.

All p not dividing $2\,\mathrm{disc}\,(F_2)$ must satisfy one of the above cases, and so we always have

$$C_2 \times C_4 \leqslant \tilde{J}(\mathbb{F}_p) \text{ or } C_2 \times C_2 \times C_2 \leqslant \tilde{J}(\mathbb{F}_p). \tag{8.2.13}$$

In all cases, $8 | \#\tilde{J}(\mathbb{F}_p)$, which means that the mere consideration of the greatest common divisor of the orders of $\tilde{J}(\mathbb{F}_p)$ would never have been sufficient to determine $\mathfrak{G}_{\mathrm{tors}}$.

The above discussion, although usually sufficient in practice, does not describe an effective procedure for finding all members of $\mathfrak{G}_{\mathrm{tors}}$. We shall see in Chapter 12 that a height function on $\mathfrak{G}$ gives a crude effective procedure, described by (12.2.20), for jacobians of both elliptic and genus 2 curves. In the case of an elliptic curve $Y^2 = (\text{cubic in } X)$ there is a much faster effective procedure given by the theorem in [Nagell (1935), Lutz (1937)], also described in [Silverman (1986), p. 221], that if (x, y) is torsion then $x, y \in \mathbb{Z}$ and $y^2 | \mathrm{disc}\,(F)$. We are not aware of anything analogous to this for jacobians of curves of genus 2.

3   Searching for large rational torsion. Consider first the elliptic curve situation. Suppose that we are looking for a parametrized family of elliptic curves defined over $\mathbb{Q}$ with a rational 5-torsion point. One approach is to write the curve in the form

$$\mathcal{E}_{A,\lambda}: \quad Y^2 = \left(A(X)\right)^2 - \lambda X(X-1)^2, \tag{8.3.1}$$

where $A(X)$ is a linear polynomial in X defined over $\mathbb{Q}(t)$, and $\lambda \in \mathbb{Q}(t)$. Let

$$P_0 = \left(0, A(0)\right), \quad P_1 = \left(1, A(1)\right). \tag{8.3.2}$$

Then the divisor of the function $Y - A(x)$ gives that

$$1 \cdot P_0 + 2 \cdot P_1 = \mathfrak{D}. \tag{8.3.3}$$

Suppose it were also true that

$$3 \cdot P_0 + 1 \cdot P_1 = \mathfrak{D}. \tag{8.3.4}$$

Then it would follow that

$$M \cdot \begin{pmatrix} P_0 \\ P_1 \end{pmatrix} = \begin{pmatrix} \mathfrak{D} \\ \mathfrak{D} \end{pmatrix}, \quad M = \begin{pmatrix} 1 & 2 \\ 3 & 1 \end{pmatrix}. \tag{8.3.5}$$

On multiplying both sides by the matrix $-\det(M) \cdot M^{-1} \in \mathbb{M}_2(\mathbb{Z})$

$$5 \cdot \begin{pmatrix} P_0 \\ P_1 \end{pmatrix} = -\det(M) \cdot \begin{pmatrix} P_0 \\ P_1 \end{pmatrix} = \begin{pmatrix} \mathfrak{D} \\ \mathfrak{D} \end{pmatrix}, \tag{8.3.6}$$

so that P_0, P_1 would both be torsion of order 5. Therefore, it is sufficient to choose A and λ so that (4) is true. We require a function $Y - v(X)$, where v is linear in X and defined over $\mathbb{Q}(t)$, whose divisor gives rise to the condition (4) on P_0, P_1. On substituting $v(X)$ for Y in (1), we require the polynomial in X, given by the difference of the two sides of (1), to be $X^3(X - 1)$. This gives

$$v^2 - A^2 + \lambda X(X - 1)^2 = X^3(X - 1), \tag{8.3.7}$$

which may be rearranged as

$$(v + A)(v - A) = X(X - 1)(X^2 - \lambda(X - 1)). \tag{8.3.8}$$

This can be directly solved by letting λ be the free parameter t, and dividing the factors of the right hand side between $v + A$ and $v - A$.

$$\begin{aligned} v - A &= X(X - 1) \\ v + A &= X^2 - \lambda(X - 1). \end{aligned} \tag{8.3.9}$$

On solving for A, λ we find

$$A = \frac{1}{2}(X^2 - t(X - 1) - X(X - 1)) = \frac{1}{2}((1 - t)X + t), \quad \lambda = t. \tag{8.3.10}$$

This gives the following family of examples over $\mathbb{Q}$.

Example 8.3.1. *For any $t \in \mathbb{Q}$, $t \neq 0$, the curve $\mathcal{E}_{A,\lambda}$ of (1) with A, λ as in (10) will have the point $P_0 \in \mathcal{E}(\mathbb{Q})$ of torsion order 5.*

As a variation, suppose now that we would like a 7-torsion point, rather than a 5-torsion point. It is sufficient to change the bottom left hand entry of the matrix M of (5) from 3 to -3, which would change the determinant of M from -5 to 7. We replace the condition (4) by

$$-3 \cdot P_0 + 1 \cdot P_1 = \mathfrak{O}. \qquad (8.3.11)$$

Then everything is the same as before up to and including the expression for $(v+A)(v-A)$ given in (8). However, the alteration from 3 to -3 means that we require $v(0) = -A(0)$, $v(0) \neq A(0)$, $v(1) = A(1)$, $v(1) \neq -A(1)$. These conditions give, respectively, that $X|(v+A)$, $X \nmid (v-A)$, $(X-1)|(v-A)$ and $(X-1) \nmid (v+A)$. Rather than assigning all of $X(X-1)$ to $v - A$ as in (9), we must write

$$v - A = Xp, \quad v + A = (X-1)q, \quad p = X - \alpha, \quad q = X - \beta, \qquad (8.3.12)$$

where $\alpha, \beta \in \mathbb{Q}(t)$. Solving for A and q gives

$$A = \big(Xp - (X-1)q\big)/2, \quad q = \big(X^2 - \lambda(X-1)\big)/p. \qquad (8.3.13)$$

Our requirement is that q be a polynomial; in other words, that p divides $\big(X^2 - \lambda(X-1)\big)$. This is equivalent to

$$\alpha^2 - \lambda(\alpha - 1) = 0, \qquad (8.3.14)$$

where $\alpha, \lambda \in \mathbb{Q}(t)$. The problem of searching for curves with 7-torsion can therefore be solved by finding a parametrization of (14). We can take $\alpha = t$, giving $p = X - t$ and

$$A = \frac{-1}{2(t-1)}\big((t^2 - 3t + 1)X + t\big), \quad \lambda = t^2/(t-1). \qquad (8.3.15)$$

Example 8.3.2. *For any $t \in \mathbb{Q}$, $t \neq 0$, the curve $\mathcal{E}_{A,\lambda}$ of (1) with A, λ as in (15) will have the point $P_0 \in \mathcal{E}(\mathbb{Q})$ of torsion order 7.*

It is well known that there are elliptic curves over $\mathbb{Q}$ with rational N-torsion for $N = 1, \ldots, 10$ and 12, and it follows from Mazur (1977) that no others can occur.

In genus 2, the same ideas can be used to search for new torsion orders which do not occur in the elliptic curve situation. Write the curve of genus 2 in the form

$$\mathcal{C}_{A,\lambda}: \quad Y^2 = \big(A(X)\big)^2 - \lambda X^2(X-1)^3, \qquad (8.3.16)$$

where $A(X)$ is a quadratic polynomial in X defined over $\mathbb{Q}(t)$, and $\lambda \in \mathbb{Q}(t)$. As usual, let $P_0 = (0, A(0))$ and $P_1 = (1, A(1))$. Then the divisor of the function $Y - A(x)$ gives that $2 \cdot P_0 + 3 \cdot P_1 = \mathfrak{O}$. Suppose it were also true that

$$5 \cdot P_0 + 1 \cdot P_1 = \mathfrak{O}. \tag{8.3.17}$$

Then it would follow that

$$M \cdot \begin{pmatrix} P_0 \\ P_1 \end{pmatrix} = \begin{pmatrix} \mathfrak{O} \\ \mathfrak{O} \end{pmatrix}, \quad M = \begin{pmatrix} 2 & 3 \\ 5 & 1 \end{pmatrix}. \tag{8.3.18}$$

As before, since -13 is the determinant of M, we would then get the 13-torsion divisor $\{P_0, \infty\}$, which is the notation of Section 1 for the divisor $P_0 - \infty$. We wish to find a function $Y - v(X)$, where v is cubic in X and defined over $\mathbb{Q}(t)$, whose divisor gives rise to the the condition (17) on P_0, P_1. On substituting $v(X)$ for Y in (16), we require the polynomial in X, given by the difference of the two sides of (16), to be $X^5(X - 1)$. We may rearrange this to obtain the equation analogous to (8).

$$(v + A)(v - A) = X^2(X - 1)\left(X^3 - \lambda(X - 1)^2\right). \tag{8.3.19}$$

As with the elliptic curve case, this can be directly solved by letting λ be the free parameter t, and dividing the factors of the right hand side between $v + A$ and $v - A$.

$$\begin{aligned} v - A &= X^2(X - 1) \\ v + A &= X^3 - \lambda(X - 1)^2. \end{aligned} \tag{8.3.20}$$

On solving for A, λ we find

$$\begin{aligned} A &= \frac{1}{2}\left(X^3 - t(X - 1)^2 - X^2(X - 1)\right) \\ &= \frac{1}{2}\left((1 - t)X^2 + 2tX - t\right), \\ \lambda &= t. \end{aligned} \tag{8.3.21}$$

This gives the following family of genus 2 examples over $\mathbb{Q}$, of which that in Flynn (1990b) is a special case, whose jacobians have a new torsion order not occurring for elliptic curves.

Example 8.3.3. For any $t \in \mathbb{Q}$, $t \neq 0$, the curve $\mathcal{C}_{A,\lambda}$ of (16) with A, λ as in (21) will have the divisor $\{P_0, \infty\} \in \mathfrak{G}$ of torsion order 13.

Suppose now that we would like a 17-torsion point, rather than a 13-torsion point. It is sufficient to change the bottom left hand entry of the matrix M of (18) from 5 to -5, which would change the determinant of M from -13 to 17. We replace the condition (17) by

$$-5 \cdot P_0 + 1 \cdot P_1 = \mathfrak{O}. \qquad (8.3.22)$$

Then the expression for $(v + A)(v - A)$ is as given in (19). However, the alteration from 5 to -5 means that we require $v(0) = -A(0)$, $v(0) \neq A(0)$, $v(1) = A(1)$ and $v(1) \neq -A(1)$. These conditions give, respectively, that $X \mid (v + A)$, $X \nmid (v - A)$, $(X - 1) \mid (v - A)$ and $(X - 1) \nmid (v + A)$. Rather than assigning all of $X^2(X - 1)$ to $v - A$ as in (20), we must write

$$v - A = X^2 p, \quad v + A = (X - 1)q, \quad p = X - \alpha, \qquad (8.3.23)$$

where $\alpha \in \mathbb{Q}(t)$ and q is a monic quadratic polynomial in X defined over $\mathbb{Q}(t)$. Solving for A and q gives

$$A = (X^2 p - (X - 1)q)/2, \quad q = (X^3 - \lambda(X - 1)^2)/p. \qquad (8.3.24)$$

Our requirement is that q be a polynomial; in other words, that p divides $(X^3 - \lambda(X - 1)^2)$. This is equivalent to

$$\alpha^3 - \lambda(\alpha - 1)^2 = 0, \qquad (8.3.25)$$

where $\alpha, \lambda \in \mathbb{Q}(t)$. We can take $\alpha = t$, giving $p = X - t$ and

$$A = \frac{-1}{2(t - 1)^2}((t^3 - 5t^2 + 4t - 1)X^2 + (3t^2 - t)X - t^2),$$
$$\lambda = t^3/(t - 1)^2. \qquad (8.3.26)$$

This gives the result observed in Leprévost (1991b).

Example 8.3.3. *For any* $t \in \mathbb{Q}$, $t \neq 0$, *the curve* $C_{A,\lambda}$ *of (16) with* A, λ *as in (26) will have the divisor* $\{P_0, \infty\} \in \mathfrak{G}$ *of torsion order 17.*

A refinement, used by Ogawa (1994), is to consider functions of the form $u(X) \cdot Y - v(X)$, where u is linear in X and v is quartic in X, and replaces $X^2(X - 1)^3$ in (16) by $X^4(X - 1)$. Using a somewhat more subtle parametrization, one can then derive curves of genus 2 over $\mathbb{Q}(t)$ whose jacobians have rational 23-torsion.

The record torsion of 29 over $\mathbb{Q}$ has been found by Leprévost (1995a), using the determinant of a 3×3 matrix. This applies to the curve

$$Y^2 = (A_1(X))^2 - 8X^3(X - 1)^2$$
$$= (A_2(X))^2 + 8X^2(X - 1)^3 \qquad (8.3.27)$$
$$= (A_3(X))^2 + 16X(X - 1)^2,$$

where

$$A_1(X) = -2X^3 - X^2 + 4X - 2,$$
$$A_2(X) = -2X^3 + 3X^2 - 4X + 2,\qquad (8.3.28)$$
$$A_3(X) = 2X^3 - X^2 - 2,$$

In this case, we take

$$D_1 = \{(0, A(0)), \infty^-\},$$
$$D_2 = \{(1, A(1)), \infty^-\},\qquad (8.3.29)$$
$$D_3 = \{\infty^-, \infty^-\}.$$

Then the functions $Y - A_i(X)$ give the three rows of the following identity:

$$\begin{pmatrix} 3 & 2 & -2 \\ -2 & 3 & 0 \\ 1 & 1 & -3 \end{pmatrix} \cdot \begin{pmatrix} D_1 \\ D_2 \\ D_3 \end{pmatrix} = \begin{pmatrix} \mathfrak{O} \\ \mathfrak{O} \\ \mathfrak{O} \end{pmatrix}. \qquad (8.3.30)$$

The determinant of this matrix is 29, so that each of D_1, D_2, D_3 gives a 29-torsion member of $\mathfrak{G}$

In the other direction, there is no known bound on $\mathbb{Q}$-rational torsion on jacobians of curves of genus 2. In the special case when the jacobian is of CM type, the results of Silverberg (1988) imply that the order of a $\mathbb{Q}$-rational torsion point is at most $185640 = 2^3 \cdot 3 \cdot 5 \cdot 7 \cdot 13 \cdot 17$.

It has been observed that some of the above curves are special cases of sequences of hyperelliptic curves whose jacobians have rational torsion orders increasing at a quadratic rate with respect to genus. For example the elliptic curve of Example 1 and the genus 2 curve of Example 2 are both special cases of the following sequence of hyperelliptic curves in Flynn (1991).

$$C^{(g)}: Y^2 = \left(A^{(g)}(X)\right)^2 - tX^g(X-1)^{g+1}, \qquad (8.3.31)$$

where

$$A^{(g)}(X) = \frac{1}{2}\left(X^{g+1} - t(X-1)^g - X^g(X-1)\right). \qquad (8.3.32)$$

Then the matrix conditions of (5) and (18) generalize to

$$M \cdot \begin{pmatrix} P_0 \\ P_1 \end{pmatrix} = \begin{pmatrix} \mathfrak{O} \\ \mathfrak{O} \end{pmatrix}, \quad M = \begin{pmatrix} g & g+1 \\ 2g+1 & 1 \end{pmatrix}. \qquad (8.3.33)$$

The torsion order of $\{P_0, \infty\}$ therefore divides $2g^2 + 2g + 1$, and it can be shown that the torsion order is in fact always equal to $2g^2 + 2g + 1$. For improved variations of this theme see Leprévost (1992, 1995b,c).

Chapter 9

The isogeny.¶ Theory

0 Introduction. The discussion in Chapter 10 is largely independent of what is done here. A reader who is primarily interested in the computational aspects need read only this introduction before moving on.

We start by recalling the situation for a curve $\mathcal{E}$ of genus 1. Here the kernel of multiplication by 2 is of order 4. Over an algebraically closed field, this can be factored into the composite of an isogeny $\mathcal{E} \to \widehat{\mathcal{E}}$ and an isogeny $\widehat{\mathcal{E}} \to \mathcal{E}$, each with kernel of order 2. Here $\widehat{\mathcal{E}}$ is another curve of genus 1, in general distinct from $\mathcal{E}$. There are three such sequences, one for each of the groups of order 2 on $\mathcal{E}$. When one of the isogenies is defined over the ground field, the investigation of the Mordell-Weil group is markedly easier than it is for a general curve of genus 1.

1 For curves $\mathcal{C}$ of genus 2, the situation is somewhat different. The kernel of multiplication by 2 on the jacobian is of order 16. It can be factored into the product of two isogenies of abelian varieties, each with kernel of order 4: but only certain subgroups of order 4 are plausible as kernels of an isogeny with another jacobian.† For such subgroups, the isogenous variety is the jacobian of a curve $\widehat{\mathcal{C}}$ which can be simply described. Over the reals, this was found by Richelot‡ (1836, 1837) and used to compute

---

¶ Note that we use hat (ˆ) for isogenous pairs. The reader who is conditioned to use it for duality is hereby warned!

† Any finite subgroup on an abelian variety can be the kernel of an isogeny of abelian varieties [Mumford (1974), Theorem 4 on p. 72]. But a jacobian has a natural principal polarization. This induces a principal polarization on the quotient variety precisely when the restriction of the 2-Weil pairing to the kernel is trivial [Milne (1986c), Proposition 16.8 on p. 135]. We are indebted to Ed Schaefer for clarifying our rather fuzzy ideas.

‡ Friedrich Julius Richelot (1808–1875) was one of the surprisingly large number of mathematicians originating in Königsberg. Jacobi's favourite pupil, he succeeded him there in the University [cf. Klein (1926)]. Is also in Poggendorf.

elliptic integrals with great rapidity and precision. Humbert (1901), as part of a wider investigation, described the situation in terms of the geometry of the conic. Bost & Mestre (1988) give a good account of all this¶ and carry it further (still over the reals).

When the Richelot isogeny is defined over the rationals, it greatly facilitates the determination of the Mordell-Weil group. That is the subject of Chapters 10, 11. In this chapter, we investigate the isogeny itself over a general field such as the rationals.

2   Let C be a curve of genus 2 with a Richelot isogeny, and suppose that all the elements of the kernel are rational.† Then C is of the form

$$C: \quad Y^2 = F(X) = G_1(X)G_2(X)G_3(X), \tag{9.2.1}$$

where the

$$G_j(X) = g_{j2}X^2 + g_{j1}X + g_{j0} \tag{9.2.2}$$

are quadratics defined over the ground field. The elements of the kernel other than the identity are the three divisors of order 2 given by $G_j = 0$. The curve with the isogenous jacobian, or, as we shall say, the isogenous curve, is

$$\widehat{C}: \quad \Delta Y^2 = H(X) = L_1(X)L_2(X)L_3(X), \tag{9.2.3}$$

where‡

$$L_1(X) = G_2'(X)G_3(X) - G_2(X)G_3'(X) \tag{9.2.4}$$

and so on, cyclically, and $\Delta = \det(g_{ij})$. One has to exclude the case $\Delta = 0$, but then the jacobian of C is the product of elliptic curves (Chapter 14). It is easy to see that the relation between C and $\widehat{C}$ is essentially symmetric.*

¶ They also reproduce an examination paper which (with hints) requires the candidate to reinvent Richelot's algorithm. It was part of the entrance examination for the *École Normale Supérieure* in 1988! For physicists!!!

† Ed Schaefer points out that there are interesting Richelot isogenies where the kernel is rational as a whole, but the elements are not. We shall not be discussing them.

‡ If the G_j are expressed as homogeneous quadratic forms in (say) X_1, X_2, then L_1 is the Jacobian (= functional determinant) of $\{G_2, G_3\}$ with respect to $\{X_1, X_2\}$.

* In particular, the analogue of Δ for $\widehat{C}$ is $2\Delta^2$, so we do not have a further condition like $\Delta \neq 0$ from $\widehat{C}$.

Bost & Mestre (1988) derive the isogeny from a 2–2 correspondence between $\mathcal{C}$ and $\widehat{\mathcal{C}}$. Express $\widehat{\mathcal{C}}$ in terms of variables (Z, T) instead of (X, Y). It is readily verified that

$$\sum_j G_j(X)L_j(Z) + (X - Z)^2 \Delta = 0. \tag{9.2.5}$$

The 2–2 correspondence¶

$$G_1(X)L_1(Z) + G_2(X)L_2(Z) = 0 \tag{9.2.6}$$

between X and Z extends to $\mathcal{C}$ and $\widehat{\mathcal{C}}$ on putting

$$YT = G_1(X)L_1(Z)(X - Z). \tag{9.2.7}$$

We, however, shall not pursue this line.

3   Given $\mathcal{C}$ and assuming, just for the moment, the existence of $\widehat{\mathcal{C}}$ with the properties just described, the isogeny from $\mathcal{C}$ to $\widehat{\mathcal{C}}$ induces an algebraic map

$$\phi : \quad \mathcal{K} \longrightarrow \widehat{\mathcal{K}} \tag{9.3.1}$$

between the corresponding Kummers. We lose nothing by going down to the Kummers, because (1) lifts automatically to a map of abelian varieties. What we shall do, is construct $\widehat{\mathcal{K}}$ and ϕ directly and study their properties.

Let $\mathfrak{d}_j$ be the element of $J(\widehat{\mathcal{C}})$ which corresponds to the divisor $G_j(X) = 0$, $Y = 0$. Then $\mathfrak{d}_j$ is of order 2, and $\mathfrak{d}_1 + \mathfrak{d}_2 + \mathfrak{d}_3 = 0$. As in Chapter 3, Section 2, addition of $\mathfrak{d}_j$ on the jacobian induces a linear projective map of $\mathcal{K}$ onto itself. This map lifts to a linear map $\mathbf{W}_j$ of the underlying affine space. A key fact is that the three $\mathbf{W}_j$ commute (Chapter 3, Section 3). Then ϕ and $\widehat{\mathcal{K}}$ may be constructed by looking at (projective) invariants in the coordinate ring under the actions of the $\mathbf{W}_j$.

We also have a description of ϕ^{-1}. This identifies the function field† $k(\widehat{\mathcal{K}})$ with a subfield of $k(\mathcal{K})$. The extension is clearly normal of relative degree 4, the automorphisms being induced by addition of the $\mathfrak{d}_j$. Hence ϕ^{-1} is described by the adjunction of two independent square roots of elements of $k(\widehat{\mathcal{K}})$.

Much good clean innocent fun will be got from reconciling the descriptions of ϕ and of ϕ^{-1}. And the results are extremely useful.

¶ Note that it is not symmetric in the index j.

† As usual, k is the ground field, say $\mathbb{Q}$.

The first thing to do, however, is to examine the additional structure on $\mathcal{K}$ that arises because $F(X)$ factors as $G_1(X)G_2(X)G_3(X)$.

4   The Kummer formulation. We shall leave the details of the transcription until later, but the main lines are clear. We have a Kummer $\mathcal{K}$ with zero node **O** and three further rational nodes such that
(i)   the three corresponding 2-divisors and **O** form a group,
(ii) the corresponding projective maps of the ambient space lift to commuting linear maps of the underlying affine space.

An alternative formulation is that the 2-Weil pairing is trivial on the group. In the classical language of Hudson (1905) the four nodes including **O** form a *Göpel tetrahedron*.

More precisely, there are three commuting linear transformations $\mathbf{S}_1$, $\mathbf{S}_2$, $\mathbf{S}_3$ (say) of the affine space which leave the Kummer invariant and three nonzero elements e_j of the ground field such that

$$\begin{aligned}
\mathbf{S}_3\mathbf{S}_2 &= \mathbf{S}_2\mathbf{S}_3 = e_1\mathbf{S}_1, \\
\mathbf{S}_1\mathbf{S}_3 &= \mathbf{S}_3\mathbf{S}_1 = e_2\mathbf{S}_2, \\
\mathbf{S}_2\mathbf{S}_1 &= \mathbf{S}_1\mathbf{S}_2 = e_3\mathbf{S}_3.
\end{aligned} \tag{9.4.1}$$

It is convenient to treat the subscripts as integers modulo 3. With this convention, (1) implies that

$$\mathbf{S}_j^2 = e_{j+1}e_{j+2}\mathbf{I}, \tag{9.4.2}$$

where $\mathbf{I}$ is the identity. Note that the e_j are not uniquely determined by the Kummer and the tetrad of nodes. In particular, we may multiply all the e_j by the same number by multiplying all the $\mathbf{S}_j$ by its inverse.

Let **O** be a set of affine coordinates for the zero node. If

$$\mathbf{O}, \ \mathbf{S}_1\mathbf{O}, \ \mathbf{S}_2\mathbf{O}, \ \mathbf{S}_3\mathbf{O} \tag{9.4.3}$$

are linearly dependent, then the corresponding jacobian is reducible (Chapter 14). We exclude this case. Otherwise, we may take (3) as a basis by writing the general vector as

$$X\mathbf{O} + Y_1\mathbf{S}_1\mathbf{O} + Y_2\mathbf{S}_2\mathbf{O} + Y_3\mathbf{S}_3\mathbf{O}. \tag{9.4.4}$$

In these coordinates $\mathbf{S}_1$ is

$$\mathbf{S}_1: \ (X, Y_1, Y_2, Y_3) \longmapsto (e_2e_3Y_1, X, e_2Y_3, e_3Y_2), \tag{9.4.5}$$

and similarly for $\mathbf{S}_2$, $\mathbf{S}_3$.

We take the equation of $\mathcal{K}$ in the shape $K = 0$, where K is a quartic form $K(X, Y_1, Y_2, Y_3)$. The individual variables occur in K only up to degree 2 because the vertices of the tetrahedron are nodes. Since $\mathcal{K}$ is invariant under the $\mathbf{S}_j$, it is easy to see that K is a linear combination of

$$U_j = X^2 Y_j^2 + e_j^2 Y_{j+1}^2 Y_{j+2}^2,$$
$$V_j = Y_{j+1} Y_{j+2}(X^2 + e_{j+1}e_{j+2}Y_j^2) + XY_j(e_{j+2}Y_{j+1}^2 + e_{j+1}Y_{j+2}^2), \quad (9.4.6)$$
$$W = XY_1 Y_2 Y_3.$$

Here j runs from 1 to 3 and, as usual, subscripts are taken modulo 3.

5   Statement of results.

THEOREM 9.5.1. *There are s_1, s_2, s_3 in the ground field such that $\mathcal{K}$ is given by $K = 0$, where*

$$K = \sum_j t_j U_j + \sum_j n_j V_j + pW, \qquad (9.5.1)$$

and

$$t_j = s_j^2 - 4e_{j+1}e_{j+2},$$
$$n_j = 4e_j s_j - 2s_{j+1}s_{j+2}, \qquad (9.5.2)$$
$$p = 4s_1 s_2 s_3 - 8e_1 e_2 e_3 - 2\sum e_j s_j^2.$$

Unexpectedly, the map from $(e_1, e_2, e_3, s_1, s_2, s_3)$ to $(t_1, t_2, t_3, n_1, n_2, n_3)$ is an involution in $\mathbb{P}^5$. Using matrices (2) can be written¶

$$\begin{bmatrix} 2t_1 & n_3 & n_2 \\ n_3 & 2t_2 & n_1 \\ n_2 & n_1 & 2t_3 \end{bmatrix} = -2\,\mathrm{adj} \begin{bmatrix} 2e_1 & s_3 & s_2 \\ s_3 & 2e_2 & s_1 \\ s_2 & s_1 & 2e_3 \end{bmatrix}. \qquad (9.5.3)$$

On taking the adjoint on both sides of (3) one gets a similar relation, but with the rôles of the two sets of six quantities interchanged.

Let

$$\Delta = 8e_1 e_2 e_3 + 2s_1 s_2 s_3 - 2\sum_j e_j s_j^2 \qquad (9.5.4)$$

¶ We are grateful to Nick Shepherd-Barron for this observation. For the background see Ein & Shepherd-Barron (1989).

be the determinant of the matrix on the left of (3). It will soon be apparent that $\Delta = 0$ only for reducible jacobians, so we will assume $\Delta \neq 0$. We have

Lemma 9.5.1. *If the e_j, s_j on the right hand side of (2) are replaced by t_j and n_j respectively, then the left hand sides are replaced by $-8\Delta e_j$ and $-8\Delta s_j$.*

THEOREM 9.5.2. *The above involution takes the Kummer $\mathcal{K}$ into its Richelot isogenous Kummer $\widehat{\mathcal{K}}$.*

Clearly the t_j, n_j occur homogeneously in (2), but it is not perhaps immediately obvious that the e_j, s_j also occur essentially homogeneously. On a change of coordinates leaving the Y_j fixed but multiplying X by a constant, the e_j and s_j are multiplied by the same constant. [For the s_j this will be more apparent when they are introduced.]

We leave the proofs of these theorems until later in the chapter.

6 Motivation. Consider a Kummer $\mathcal{K}$ invariant under linear transformations $\mathbf{S}_j$ satisfying (4.1). By the properties of incidence there are precisely two tropes through any two given nodes. We may take the dual coordinates of one of the tropes through $(0,0,1,0)$ and $(0,0,0,1)$ in the shape $(1, -c_1, 0, 0)$, where c_1 is not in general in the ground field. The automorphism $\mathbf{S}_1$ interchanges the two nodes and takes the trope into $(-c_1, e_2e_3, 0, 0)$, which must be the other trope through the two nodes. The two tropes are rational¶ as a pair, so

$$c_1^2 - s_1 c_1 + e_2 e_3 = 0 \qquad (9.6.1)$$

for some rational s_1. The individual tropes are defined over $k(\tau_1)$, where $\tau_1^2 = s_1^2 - 4e_2e_3 = t_1$ in the notation of (5.2). On applying $\mathbf{S}_2$ or $\mathbf{S}_3$ we get the two tropes through $(1,0,0,0)$ and $(0,1,0,0)$. Similarly for the other pairs of vertices of the tetrahedron of reference.

We now have 12 of the tropes. Their intersections give the full complement of 16 nodes. The nodes determine $\mathcal{K}$, so we could, if we wanted,† compute its equation in terms of the e_j, s_j.

There are four tropes which do not meet the vertices of the tetrahedron of reference. They are clearly each defined over $k(\tau_1, \tau_2, \tau_3)$ and rational as a set. Since there are only four of them, they must be defined over

---

¶ meaning defined over the ground field k

† We suppose. Courage failed us.

$k(\tau_2\tau_3, \tau_3\tau_1, \tau_1\tau_2)$. Each node, other than the vertices of the tetrahedron of reference, lies on precisely two of these tropes. By general theory, there is a quadric through the 12 nodes. It is unique, so given by $Q = 0$, where Q is a rational quadratic form. Let L_j $(j = 1, \ldots, 4)$ be conjugate linear forms in X, Y_1, Y_2, Y_3 giving the tropes. Then Q^2, $L_1L_2L_3L_4$ and the quartic form K giving $\mathcal{K}$ are linearly dependent, say without loss of generality

$$K = Q^2 - \rho L_1 L_2 L_3 L_4 \tag{9.6.2}$$

for some rational ρ.

There is another quadratic form and set of four conjugate linear forms associated with the configuration. Put $\epsilon_j^2 = e_j$ and write

$$\begin{aligned}
X &= \epsilon_1\epsilon_2\epsilon_3(v_0 + v_1 + v_2 + v_3), \\
Y_1 &= \epsilon_1(v_0 + v_1 - v_2 - v_3), \\
Y_2 &= \epsilon_2(v_0 - v_1 + v_2 - v_3), \\
Y_3 &= \epsilon_3(v_0 - v_1 - v_2 + v_3).
\end{aligned} \tag{9.6.3}$$

Here v_0, v_1, v_2, v_3 are linear forms in X, Y_1, Y_2, Y_3. Each transformation $\mathbf{S}_j/\epsilon_{j+1}\epsilon_{j+2}$ changes the sign of precisely two of the v_k: its square is the identity $\mathbf{I}$. But K is invariant under these transformations, so

$$K = q(v_0^2, v_1^2, v_2^2, v_3^2) + \gamma v_0 v_1 v_2 v_3, \tag{9.6.4}$$

where q is a quadratic form and γ is a constant. Hence $K = 0$ implies that

$$q(V_0, V_1, V_2, V_3)^2 - \gamma^2 V_0 V_1 V_2 V_3 = 0, \tag{9.6.5}$$

where $V_j = v_j^2$.

The quartic surface (5) in the variables V_j is the quotient of $\mathcal{K}$ by the $\mathbf{S}_j$, and so must be the isogenous $\widehat{\mathcal{K}}$. As it stands, it is not defined over the ground field. But the 16 nodes of $\mathcal{K}$ map by fours into the special four nodes of $\widehat{\mathcal{K}}$. On taking these as basis of coordinates, we get a map defined over the ground field. The analogy between (5) and (2) is apparent.

In what follows, we shall not go into the details of the proof of Theorem 5.1 as outlined above: we should need to assume either a preternatural hindsight or some heroic algebra. We shall, however, sketch the construction of the isogenous Kummer in some detail. The resulting insight should enable the reader to construct a tolerable proof of Theorem 5.1 if she wishes.

7   The isogeny. In this section we implement the program of the previous section and construct the isogeny. We start with $\mathcal{K}$ given by Theorem 5.1 and with the v_j, V_j, q, γ of (6.3–5). Our first goal is to eliminate the ϵ_j, so everything will be defined over the ground field. We have

$$
\begin{aligned}
V_0 &= M_0 + 2\epsilon_2\epsilon_3 M_1 + 2\epsilon_3\epsilon_1 M_2 + 2\epsilon_1\epsilon_2 M_3,\\
V_1 &= + - - ,\\
V_2 &= - + - ,\\
V_3 &= - - + ,
\end{aligned}
\tag{9.7.1}
$$

where

$$
\begin{aligned}
M_0 &= X^2 + e_2 e_3 Y_1^2 + e_3 e_1 Y_2^2 + e_1 e_2 Y_3^2,\\
M_1 &= XY_1 + e_1 Y_2 Y_3,\\
M_2 &= XY_2 + e_2 Y_3 Y_1,\\
M_3 &= XY_3 + e_3 Y_1 Y_2.
\end{aligned}
\tag{9.7.2}
$$

It remains to verify that we actually have a Kummer surface and to bring its equation to canonical form. This requires the relations (5.2). The 16 nodes of $\mathcal{K}$ map by fours into the four special rational nodes of $\widehat{\mathcal{K}}$. We suppress the calculations, which suggest taking new coordinates x and y_j given by

$$
\begin{aligned}
M_0 &= x + s_2 s_3 y_1 + s_3 s_1 y_2 + s_1 s_2 y_3,\\
M_1 &= 2e_1 y_1 + s_3 y_2 + s_2 y_3,\\
M_2 &= s_3 y_1 + 2e_3 y_2 + s_1 y_3,\\
M_3 &= s_2 y_1 + s_1 y_2 + 2e_3 y_3.
\end{aligned}
\tag{9.7.3}
$$

On substituting in (6.5) and removing a scalar factor we obtain the surface $\widehat{\mathcal{K}}$ required by Theorem 5.2 in coordinates x and y_j.

Noting that the matrix of the last three lines of (3) occurs in (5.3), we can solve for the y_j and x:

$$
\begin{aligned}
y_1 &= 2t_1(XY_1 + e_1 Y_2 Y_3) + n_3(XY_2 + e_2 Y_1 Y_3) + n_2(XY_3 + e_3 Y_1 Y_2),\\
y_2 &= n_3(XY_1 + e_1 Y_2 Y_3) + 2t_2(XY_2 + e_2 Y_1 Y_3) + n_1(XY_3 + e_3 Y_1 Y_2),\\
y_3 &= n_2(XY_1 + e_1 Y_2 Y_3) + n_1(XY_2 + e_2 Y_1 Y_3) + 2t_3(XY_3 + e_3 Y_1 Y_2).\\
x &= -2\Delta(X^2 + e_2 e_3 Y_1^2 + e_3 e_1 Y_2^2 + e_1 e_2 Y_3^2),
\end{aligned}
\tag{9.7.4}
$$

8   **Generation by radicals.** In the previous sections we constructed the isogenous Kummer $\widehat{\mathcal{K}}$ using the coordinate ring $k[\mathcal{K}]$, that is the quotient of $k[X, Y_1, Y_2, Y_3]$ by the defining polynomial $K(X, Y_1, Y_2, Y_3)$. As always, k is the ground field. The action of the $\mathbf{S}_j$ is more pleasant if we restrict ourselves to a subring.

Let $k[\mathcal{K}]^{[2]}$ be the subring of elements of even degree of $k[\mathcal{K}]$. It is generated over k by the space $\mathbf{V}$ (say) of forms of degree 2. The k-linear maps

$$\mathbf{H}_j(v) = \mathbf{S}_j(v)/(e_{j+1}e_{j+2}), \qquad j = 1, 2, 3 \qquad v \in \mathbf{V} \qquad (9.8.1)$$

satisfy

$$\mathbf{H}_j^2 = \mathbf{I}, \qquad \mathbf{H}_i\mathbf{H}_j = \mathbf{H}_j\mathbf{H}_i, \qquad \mathbf{H}_1\mathbf{H}_2\mathbf{H}_3 = \mathbf{I}. \qquad (9.8.2)$$

The space $\mathbf{V}$ splits into four eigenspaces, $\mathbf{V}_j$:

$\mathbf{V}_0$ of dimension 4 of elements invariant under all the $\mathbf{H}_j$. It is spanned by the $XY_j + e_j Y_{j+1} Y_{j+2}$ $(j = 1, 2, 3)$ and $X^2 + e_2 e_3 Y_1^2 + e_3 e_1 Y_2^2 + e_1 e_2 Y_3^2$.

$\mathbf{V}_j$ $(j = 1, 2, 3)$ of dimension 2 spanned by $XY_j - e_j Y_{j+1} Y_{j+2}$ and $X^2 + e_{j+1} e_{j+2} Y_j^2 - e_{j+2} e_j Y_{j+1}^2 - e_j e_{j+1} Y_{j+2}^2$.

The right hand sides of (7.4) span¶ $\mathbf{V}_0$. The equations (7.4) extend to an embedding of $k[\widehat{\mathcal{K}}]$ in $k[\mathcal{K}]^{[2]}$: we identify $k[\widehat{\mathcal{K}}]$ with its image. The $\mathbf{H}_j$ extend in the obvious way to $k[\mathcal{K}]^{[2]}$. Clearly $k[\widehat{\mathcal{K}}]$ is elementwise invariant, but an invariant element† need not be in $k[\widehat{\mathcal{K}}]$.

The squares of elements of the $\mathbf{V}_j$ $(j \neq 0)$ are invariant under the $\mathbf{H}_j$. We want to find such a square which is in $k[\widehat{\mathcal{K}}]$. Equations of tropes naturally suggest themselves because their divisors of zeros are divisible by 2. In fact

$$
\begin{aligned}
& x^2 - n_1 x y_1 + t_2 t_3 y_1^2 \\
&= 4\Delta^2 \big((X^2 - s_1 X Y_1 + e_2 e_3 Y_1^2) - e_1 (e_3 Y_2^2 - s_1 Y_2 Y_3 + e_2 Y_3^2) \big)^2
\end{aligned}
\qquad (9.8.3)
$$

in $k[\mathcal{K}]^{[2]}$ by Littlewood's Principle.‡ Here the left-hand side of (3) equated to 0 is the equation of a rational pair of tropes of $\widehat{\mathcal{K}}$.

Recall that the function field $k(\mathcal{K})$ consists of the quotients of homogeneous elements of $k[\mathcal{K}]$ of the same degree, which is even without loss of generality. Putting everything together, we have proved

---

¶ Remember that $\Delta \neq 0$.

† By abstract nonsense it is in the integral closure. Cf. Chapter 16.

‡ 'All identities are trivial' (it being understood that they have been already written down by someone else).

THEOREM 9.8.1. $k(\mathcal{K})/k(\widehat{\mathcal{K}})$ *is normal of degree 4 and type* $(2,2)$. *The extension is generated redundantly by* ζ_j $(j = 1, 2, 3)$, *where*

$$x^2 \zeta_j^2 = x^2 - n_j x y_j + t_{j+1} t_{j+2} y_j^2. \tag{9.8.4}$$

The nontrivial automorphisms τ_j $(j = 1, 2, 3)$ *can be labelled so that*

$$\tau_i \zeta_j = \begin{cases} \zeta_j & \text{if } j = i, \\ -\zeta_j & \text{otherwise.} \end{cases} \tag{9.8.5}$$

9   Consequences for Mordell-Weil. Suppose, now, that $\mathcal{K}$ comes from an abelian variety $\mathcal{A}$ and let $\widehat{\mathcal{A}}$ be obtained by factoring out the special subgroup of order 4. Then $\widehat{\mathcal{K}}$ is the Kummer of $\widehat{\mathcal{A}}$. Let $\mathfrak{G}$ be the group of rational points on $\mathcal{A}$ (the Mordell-Weil group), and similarly for $\widehat{\mathfrak{G}}$. The map from $\mathcal{A}$ to $\widehat{\mathcal{A}}$ induces a map from $\mathfrak{G}$ to $\widehat{\mathfrak{G}}$. Denote the image by im $\mathfrak{G}$. In this section we give a description of the quotient $\widehat{\mathfrak{G}}/\text{im }\mathfrak{G}$ reminiscent of the description of $\mathfrak{G}/2\mathfrak{G}$ of Chapter 6.

In the notation of Theorem 8.1 put

$$\mathbf{Z}_j = (x^2 - n_j x y_j + t_{j+1} t_{j+2} y_j^2)/x^2 \tag{9.9.1}$$

regarded as a function on $\widehat{\mathcal{A}}$. Let $\mathfrak{a} \in \widehat{\mathfrak{G}}$. If $\mathfrak{a}$ is neither a zero nor a pole of $\mathbf{Z}_j$, let $\mu_j(\mathfrak{a})$ be the class of $\mathbf{Z}_j(\mathfrak{a})$ in k^*/k^{*2}. If $\mathfrak{a}$ is a zero or pole, there is a function Φ on $\widehat{\mathcal{A}}$ such that $\mathfrak{a}$ is neither a zero nor a pole of $\mathbf{Z}_j \Phi^2$, and we use this instead of $\mathbf{Z}_j$. Put

$$\mu(\mathfrak{a}) = \big(\mu_1(\mathfrak{a}), \mu_2(\mathfrak{a}), \mu_3(\mathfrak{a})\big). \tag{9.9.2}$$

THEOREM 9.9.1. *The map* $\widehat{\mathfrak{G}} \to (k^*/k^{*2})^3$ *given by* μ *is a group homomorphism with kernel* im $\mathfrak{G}$.

Proof. Note first that

$$\mu_1(\mathfrak{a})\mu_2(\mathfrak{a})\mu_3(\mathfrak{a}) = 1 \in k^*/k^{*2} \tag{9.9.3}$$

by Theorem 8.1.

If $\mu(\mathfrak{a}) = 1$, then $\mathfrak{a} \in \text{im }\mathfrak{G}$ and conversely, by Theorem 8.1. More generally, $\mathfrak{a}$ is the image of a point $\mathfrak{A}$ (say) on $\mathcal{A}$ defined over the field extension given by the square roots of the $\mu_j(\mathfrak{a})$. Hence we have a map from $\widehat{\mathfrak{G}}$ to $H^1(\mathrm{Gal}\,(\bar{k}/k), \ker \mathfrak{G})$, which is clearly a homomorphism. But the Galois group acts trivially on the kernel $\ker \mathfrak{G}$ of the isogeny, so the cohomology group is canonically isomorphic to $(k^*/k^{*2})^2$, as required.

10   Isogeny for the curve. We apply the results just obtained to the jacobian of the curve

$$\mathcal{C}:\quad Y^2 = F(X) = G_1(X)G_2(X)G_3(X), \qquad (9.10.1)$$

where the

$$G_j(X) = g_{j2}X^2 + g_{j1}X + g_{j0} \qquad (9.10.2)$$

are defined over the ground field. In the standard coordinate system for the Kummer, the coordinates of the nodes corresponding to the G_j are

$$(g_{j2}, -g_{j1}, g_{j0}, b_j), \qquad (9.10.3)$$

where

$$b_j = -g_{j2}^2 g_{i0}g_{k0} - (g_{i1}g_{k1} + g_{i0}g_{k2} + g_{i1}g_{k0})g_{j2}g_{j0} - g_{j0}^2 g_{i2}g_{k2} \qquad (9.10.4)$$

is the unique value of the fourth coordinate giving a point on the Kummer. In this section we use the convention that $\{i, j, k\}$ is a permutation of $\{1, 2, 3\}$.

It turns out to be convenient to take the affine coordinates of **O** as $(0, 0, 0, -1)$. With these vectors as the basis of the coordinate system (X, Y_1, Y_2, Y_3), the polynomial K defining the Kummer (3.1.8) takes the shape

$$K = \sum_j T_j U_j + \sum_j N_j V_j + PW. \qquad (9.10.5)$$

Here the polynomials U_j, V_j, W are given by (4.6) with parameters e_j yet to be described, and T_j, N_j, P are coefficients also to be described.

It turns out that T_j is the discriminant of G_j and that N_j is the corresponding mixed invariant of the other two forms G_i, G_j:

$$T_j = g_{j1}^2 - 4g_{j0}g_{j2}, \qquad (9.10.6)$$
$$N_j = 2g_{i1}g_{k1} - 4g_{i0}g_{k2} - 4g_{k0}g_{i2}.$$

We shall not use the value of P. The values of the e_j and the s_j may also be read off from (5). It turns out (cf. (5.2) and Lemma 5.1) that

$$e_j = (N_j^2 - 4T_{j+1}T_{j+2})/16, \qquad (9.10.7)$$
$$s_j = (4T_j N_j - 2N_{j+1}N_{j+2})/16.$$

11   We next compare the isogeny for Kummers with the isogeny for curves. Put

$$L_j(X) = G'_{j+1}(X)G_{j+2}(X) - G'_{j+2}(X)G_{j+1}(X) \qquad (9.11.1)$$

in the notation of Section 2. It is a straightforward calculation that the discriminant of L_j is $4e_j$ and that the mixed invariant of L_{j+1}, L_{j+2} is $4s_j$: in complete conformity with the duality of Section 5.

Computation with the Kummer alone cannot confirm the constant Δ on the left hand side of (2.5), since that is not reflected in the geometry of the Kummer. To check it, one would have to lift the Kummer $\mathcal{K}$ to an abelian variety $\mathcal{A}$ and look at the effect of the isogeny on $\mathcal{A}$. Since this requires only a quadratic extension of the function-field (Chapter 3, Section 8), this is straightforward in theory. It looks as though it would be tedious in practice, and we have not done it.

12   In conclusion, we compare the map on the Mordell-Weil group obtained in Section 9 with the map obtained for general curves of genus 2 earlier. For simplicity, we assume that the three quadratics in (10.1) are irreducible. We first reformulate the map of Chapter 6 for this special situation. Let θ_j be a root of G_j. The image of the map is in

$$\prod_j \left(k(\theta_j)^* / \left(k(\theta_j)^{*2}\right) \right) \Big/ k^*. \qquad (9.12.1)$$

Here k is the ground field, regarded as a subfield of each of the $k(\theta_j)$.

A rational divisor $\mathfrak{X} = \{(x,y),(u,v)\}$ is mapped into (1) by taking $(\theta_j - x)(\theta_j - u) \in k^*(\theta_j)$. On taking the norms in the three quadratic extensions, we have a much simpler map. The image is in the product of three copies of $k^*/(k^*)^2$. The three components of the image of $\mathfrak{X}$ are just the resultants of $(X - x)(X - u)$ with the G_j. We show that the map just constructed is identical with that of Section 9, but in making the comparison the reader should note that Section 9 works with $\widehat{\mathcal{K}}$, not $\mathcal{K}$.

On the corresponding Kummer, the locus of zeros of the resultant of $(X - x)(X - u)$ and G_1 is just a pair of tropes, conjugate over the ground field. It can therefore be no surprise that in the X, Y_1, Y_2, Y_3 coordinates the resultant becomes just $e_3 Y_2^2 - s_1 Y_2 Y_3 + e_2 Y_3^2$. The equivalence of the map just obtained with that of Section 9 is an immediate consequence of

Lemma 9.12.1. *Let $\mathcal{K}$ be a Kummer in the shape (5.1) with*

$$e_j = n_j^2 - 4t_{j+1}t_{j+2} \qquad\qquad s_j = 4t_j n_j - 2n_{j+1}n_{j+2}. \qquad (9.12.2)$$

Then

$$\left(2t_1XY_1 + (n_3Y_2 + n_2Y_3)X + (n_2e_3Y_2 + n_3e_2Y_3)Y_1 + 2t_1e_1Y_2Y_3\right)^2$$
$$= (e_3Y_2^2 - s_1Y_2Y_3 + e_2Y_3^2)(X^2 - s_1XY_1 + e_2e_3Y_1^2). \qquad (9.12.3)$$

Note that the choice of the sign of X in Section 10 was to ensure that (2) holds. The Lemma follows a standard pattern. The right hand side is the product of four tropes: the left hand side a quadric taken twice. Trivial, by Littlewood's Principle again.

Chapter 10

The isogeny. Applications

0   Introduction. We give an alternative description of Richelot's isogeny as a map defined by quadratic forms on the jacobian variety in $\mathbb{P}^{15}$. This is analogous to a similar map on the $\mathbb{P}^3$ embedding of an elliptic curve. In either case, we give two alternative descriptions of an injection on $\widehat{\mathfrak{G}}/\phi(\mathfrak{G})$ which are used for performing descent via isogeny. We outline how the homogeneous spaces are constructed, and give a 'norm' space satisfying the Hasse principle, relating to each homogeneous space. We shall use elliptic curves as motivation and then describe what happens in genus 2. Finally, we show a local result which can be used to find $\#\widehat{\mathfrak{G}}/\phi(\mathfrak{G})\cdot\#\mathfrak{G}/\hat{\phi}(\widehat{\mathfrak{G}})$, when the ground field is $\mathbb{Q}_p$.

1   The isogeny on the jacobian variety.

(i) Elliptic curves. An elliptic curve over k with a rational point of order 2 may be written in the form

$$\mathcal{E}: \quad Y^2 = X(X^2 + aX + b), \ a, b \in k, \ b(a^2 - 4b) \neq 0. \qquad (10.1.1)$$

We shall as usual let z_0, z_1, z_2, z_3 be as in (7.1.2); that is $X^2, Y, X, 1$, with $J = J(\mathcal{E})$ denoting the projective locus of (z_i) in $\mathbb{P}^3$. The defining equations are now

$$z_0 z_2 = z_1^2 - a z_3^2 - b z_2 z_3$$
$$z_0 z_3 = z_2^2. \qquad (10.1.2)$$

Addition by the point $(0, 0)$ is given by a linear map W on J:

$$W = \begin{pmatrix} 0 & 0 & 0 & 1 \\ 0 & b & 0 & 0 \\ 0 & 0 & -b & 0 \\ b^2 & 0 & 0 & 0 \end{pmatrix}. \qquad (10.1.3)$$

The matrix W has eigenvalues b, $-b$, each occurring with multiplicity 2. We therefore perform a change of basis to new functions v_0, v_1, v_2, v_3 given by

$$(v_i) = M \cdot (z_i), \quad M = \begin{pmatrix} 1 & 0 & 0 & b \\ 0 & 0 & 1 & 0 \\ 1 & 0 & 0 & -b \\ 0 & 1 & 0 & 0 \end{pmatrix}, \tag{10.1.4}$$

so that W becomes diagonalized as

$$b \begin{pmatrix} I_2 & 0 \\ 0 & -I_2 \end{pmatrix}, \tag{10.1.5}$$

where I_2 is the 2×2 identity matrix. The defining equations become

$$\begin{aligned} v_2^2 &= v_0^2 - 4bv_1^2 \\ v_3^2 &= v_0 v_1 + a v_1^2. \end{aligned} \tag{10.1.6}$$

The quadratic monomials v_0^2, $v_0 v_1$, v_2^2 and $v_2 v_3$ span those which are invariant under T. These require only a further linear adjustment to give an isogeny to the jacobian $\widehat{J} = J(\widehat{\mathcal{E}})$ of the elliptic curve

$$\widehat{\mathcal{E}}: \quad Y^2 = X(X^2 + \hat{a}X + \hat{b}), \quad \hat{a} = -2a, \hat{b} = a^2 - 4b. \tag{10.1.7}$$

The 2-isogeny ϕ from J to $\widehat{J}$ may be completely described by

$$\phi = M^\star \rho M, \quad M^\star = \begin{pmatrix} b(a^2 + 4b) & 8ab^2 & -a^2 b^2 & 0 \\ 0 & 0 & 0 & 4b^2 \\ ab & 4b^2 & -ab & 0 \\ a^2 - 4b & 0 & -a^2 + 4b & 0 \end{pmatrix}, \tag{10.1.8}$$

where

$$\rho : (v_i) \longmapsto (v_i^\star) = \begin{pmatrix} v_0^2 \\ v_0 v_1 \\ v_2^2 \\ v_2 v_3 \end{pmatrix}. \tag{10.1.9}$$

(ii) Genus 2. As in Chapter 9, let

$$\begin{aligned} \mathcal{C}: \quad & Y^2 = F(X) = G_1(X)G_2(X)G_3(X), \\ \widehat{\mathcal{C}}: \quad & \Delta Y^2 = H(X) = L_1(X)L_2(X)L_3(X), \end{aligned} \tag{10.1.10}$$

where the G_i, L_i, Δ are as in Chapter 9, Section 2. Further define the quantities

$$\begin{aligned} b_{ij} &= \text{resultant}(G_i(X), G_j(X)), \quad b_i = b_{ij}b_{ik}, \\ \hat{b}_{ij} &= \text{resultant}(L_i(X), L_j(X)), \quad \hat{b}_i = \hat{b}_{ij}\hat{b}_{ik}. \end{aligned} \tag{10.1.11}$$

Let $J = J(\mathcal{C})$ be the jacobian of the curve $\mathcal{C}$ embedded into $\mathbb{P}^{15}$ by the coordinates $z_0, \ldots, z_{15}$ described at the beginning of Chapter 2, Section 2; similarly define $\widehat{J} = J(\widehat{\mathcal{C}})$. The defining equations are given by 72 quadratic forms available by anonymous ftp, as described in Appendix II. Then, as in Chapter 9, we consider the maps $\mathbf{W}_j$ for $j = 1, 2, 3$ on J. The corresponding projective maps of the ambient space lift to commuting linear maps of the underlying affine space, each of which can be represented by a 16×16 matrix defined over k. These are analogous to the 4×4 matrix W for elliptic curves given in (3).

We can now change basis to $(v_i) = M \cdot (z_i)$, where M is a 16×16 matrix defined over k, so that $\mathbf{W}_1, \mathbf{W}_2, \mathbf{W}_3$ become simultaneously diagonalized as

$$
b_1 \begin{pmatrix} I_4 & 0 & 0 & 0 \\ 0 & I_4 & 0 & 0 \\ 0 & 0 & -I_4 & 0 \\ 0 & 0 & 0 & -I_4 \end{pmatrix}, \quad b_2 \begin{pmatrix} I_4 & 0 & 0 & 0 \\ 0 & -I_4 & 0 & 0 \\ 0 & 0 & I_4 & 0 \\ 0 & 0 & 0 & -I_4 \end{pmatrix},
$$

$$
b_3 \begin{pmatrix} I_4 & 0 & 0 & 0 \\ 0 & -I_4 & 0 & 0 \\ 0 & 0 & -I_4 & 0 \\ 0 & 0 & 0 & I_4 \end{pmatrix}, \tag{10.1.12}
$$

respectively, where I_4 represents the 4×4 identity matrix. The behaviour of each v_i is the same for i within each of the sets $\{0, 1, 2, 3\}$, $\{4, 5, 6, 7\}$, $\{8, 9, 10, 11\}$, $\{12, 13, 14, 15\}$. The quadratic monomials $v_i v_j$ are invariant under all of $\mathbf{W}_1, \mathbf{W}_2$ and $\mathbf{W}_3$ whenever i and j are both in the same one of these sets. There are 40 such monomials, but the defining equations in $v_0, \ldots, v_{15}$ mean that they are all spanned by the 16 choices $v_0^\star, \ldots, v_{15}^\star$ given in the next equation.

There exists a matrix $M^\star$ such that Richelot's isogeny ϕ from J to $\widehat{J}$ can be written in the form

$$
\phi = M^\star \rho M, \quad \rho : (v_i) \longmapsto (v_i^\star), \quad v_i^\star = \begin{cases} v_0 v_i, & i = 0, \ldots, 3, \\ v_4 v_i, & i = 4, \ldots, 7, \\ v_8 v_i, & i = 8, \ldots, 11, \\ v_{12} v_i, & i = 12, \ldots, 15. \end{cases} \tag{10.1.13}
$$

This construction of the isogeny on the jacobian variety is described in more detail in Flynn (1994). It has the disadvantage that the equations are rather large, but it is at least independent of the miracle — which seems

to be special to genus 1, genus 2 and potentially genus 3¶ — that there is
a curve of which the isogenous variety is the jacobian. It also, as we shall
see in Section 3, provides a foundation for the construction of homogeneous
spaces.

2   An injection on $\widehat{\mathfrak{G}}/\phi(\mathfrak{G})$.

As usual we let $\mathfrak{G}$ denote $J(k)$ and
$\widehat{\mathfrak{G}}$ denote $\widehat{J}(k)$, whether J is either an elliptic curve or the jacobian of a
curve of genus 2. We shall not require the structure of the abelian variety
in $\mathbb{P}^{4^g-1}$ in this section, and so we shall represent a point in $\mathfrak{G}$ just as (x,y)
for the elliptic curve case, and as $\{(x,y),(u,v)\}$ for the jacobian of a curve
of genus 2.

(i) Elliptic curves. Let $\mathcal{E},\widehat{\mathcal{E}}$ be the elliptic curves in (1.1) and (1.7),
respectively. Let (x,y) be a point in $\widehat{\mathfrak{G}}$. Then there will be precisely two
points in $\mathcal{E}(\bar{k})$ which map to (x,y) under ϕ. Since ϕ is defined over k
these two points must be either both defined over k, or defined over some
quadratic extension $k(\sqrt{d})$ and conjugate. The kernel of ϕ consists of $\mathfrak{O}$
and $(0,0)$, so that $\mathbf{W}$, addition by $(0,0)$, has the same effect as conjugation
on these two points. We can define the following map:

$$\mu^\phi: \quad \widehat{\mathfrak{G}} \longrightarrow k^*/(k^*)^2, \quad (x,y) \longmapsto d, \tag{10.2.1}$$

where, for any $P \in \phi^{-1}\big((x,y)\big)$,

$$P \text{ is defined over } k(\sqrt{d}), \ \{P, \mathbf{W}(P)\} \text{ is defined over } k. \tag{10.2.2}$$

Suppose now that $(x,y) \mapsto d$ and $(x^\circ, y^\circ) \mapsto d^\circ$. Let $P \in \phi^{-1}\big((x,y)\big)$ and
$P^\circ \in \phi^{-1}\big((x^\circ, y^\circ)\big)$. Then $P + P^\circ \in \phi^{-1}\big((x,y) + (x^\circ, y^\circ)\big)$ and is defined
over $k(\sqrt{d}, \sqrt{d^\circ})$. But the action $\sqrt{d} \mapsto -\sqrt{d}$, $\sqrt{d^\circ} \mapsto -\sqrt{d^\circ}$ takes $P + P^\circ$
to $\big(P + (0,0)\big) + \big(P^\circ + (0,0)\big) = P + P^\circ$. It follows that $P + P^\circ$ is defined
over $k(\sqrt{dd^\circ})$, so that

$$\mu^\phi\big((x,y) + (x^\circ, y^\circ)\big) = dd^\circ = \mu^\phi\big((x,y)\big)\mu^\phi\big((x^\circ, y^\circ)\big). \tag{10.2.3}$$

¶ It was pointed out to us by Ed Schaefer and Joseph Wetherell that,
for a curve of genus 3, the variety isogenous to the jacobian is a principally
polarized abelian variety of dimension 3 defined over the ground field. Using
Oort and Ueno (1973), this must be the jacobian of some curve defined
over a finite extension of the ground field; we are not aware of any general
equation for this 'isogenous' curve. In genus > 3 there seems to be no reason
to expect an underlying 'isogenous' curve.

It is trivial that $\mu^\phi\big(-(x,y)\big) = d = d^{-1}$ and that $\mu^\phi(\mathfrak{O}) = 1$ which, together with (3), give that μ^ϕ is a group homomorphism. Further, (x,y) is in the kernel of μ^ϕ precisely when $P \in \mathfrak{G}$. Therefore, the induced map

$$\mu^\phi : \ \widehat{\mathfrak{G}}/\phi(\mathfrak{G}) \longrightarrow k^*/(k^*)^2, \quad (x,y) \longmapsto d, \qquad (10.2.4)$$

is injective. A bit of algebra shows that it is equivalent to define

$$\mu^\phi : \ \widehat{\mathfrak{G}}/\phi(\mathfrak{G}) \longrightarrow k^*/(k^*)^2, \quad (x,y) \longmapsto x. \qquad (10.2.5)$$

The description in (4) is better for proving that μ^ϕ is a homomorphism, whereas (5) is better for computational purposes. If we wish, we can use the equivalent

$$\mu^\phi_{\text{ext}} : \ \widehat{\mathfrak{G}}/\phi(\mathfrak{G}) \longrightarrow k^*/(k^*)^2 \times k^*/(k^*)^2,$$
$$(x,y) \longmapsto [x, x^2 + \hat{a}x + \hat{b}]. \qquad (10.2.6)$$

In this case the product of the two coordinates in the image is $y^2 \in (k^*)^2$, so that the second coordinate is redundant. However, it provides a useful device for computing μ^ϕ in the case when $(x,y) = (0,0)$; the first coordinate in (6) is undefined, and we can use the second coordinate to see that the correct image under μ^ϕ should be $\hat{b}$. This can also be seen directly from the version given in (1), (2), since the preimages of $(0,0)$ under ϕ are the points $((-b \pm \sqrt{\hat{b}})/2, 0)$, which are defined over $\mathbb{Q}(\sqrt{\hat{b}})$. If we also apply the same argument to $\mu^{\hat{\phi}}$, the corresponding injection on $\mathfrak{G}/\hat{\phi}(\widehat{\mathfrak{G}})$, then

$$\{1, \hat{b}\} \leqslant \operatorname{im} \mu^\phi$$
$$\{1, b\} \leqslant \operatorname{im} \mu^{\hat{\phi}} \qquad (10.2.7)$$

In the notation of Section 1, we can also think of $\mu^\phi\big((x,y)\big)$ as being $v_2^\star/v_0^\star$, where $v_0^\star$ and $v_2^\star$ are as in (1.8).

(ii) Genus 2. Now let $\mathcal{C}$, $\widehat{\mathcal{C}}$ be the curves of genus 2 given in (1.10), with jacobians J, $\widehat{J}$ respectively and with $\mathfrak{G} = J(k)$, $\widehat{\mathfrak{G}} = \widehat{J}(k)$. Let $\{(x,y),(u,v)\} \in \widehat{\mathfrak{G}}$. Then there will be precisely four points in $J(\bar{k})$ which map to $\{(x,y),(u,v)\}$ under Richelot's isogeny, ϕ. Since ϕ is defined over k these four points must be defined over some $k(\sqrt{d_1}, \sqrt{d_2})$, with the maps $\mathbf{W}_i$ (addition by the members of the kernel of ϕ) having the same effect as the conjugation maps on these four points. We can define the following map.

$$\mu^\phi : \ \widehat{\mathfrak{G}} \longrightarrow k^*/(k^*)^2 \times k^*/(k^*)^2, \quad (x,y) \longmapsto [d_1, d_2], \qquad (10.2.8)$$

where, for any $\mathfrak{P} \in \phi^{-1}(\{(x,y),(u,v)\})$,

$$\mathfrak{P} \text{ is defined over } k(\sqrt{d_1}, \sqrt{d_2}),$$
$$\{\mathfrak{P}, \mathbf{W}_i(\mathfrak{P})\} \text{ is defined over } k(\sqrt{d_i}), \qquad (10.2.9)$$
$$\{\mathfrak{P}, \mathbf{W}_1(\mathfrak{P}), \mathbf{W}_2(\mathfrak{P}), \mathbf{W}_3(\mathfrak{P})\} \text{ is defined over } k,$$

for $i = 1, 2, 3$, where we define $d_3 = d_1 d_2$.

Suppose $\{(x,y),(u,v)\} \mapsto [d_1, d_2]$ and $\{(x^\circ, y^\circ), (u^\circ, v^\circ)\} \mapsto [d_1^\circ, d_2^\circ]$. Let $\mathfrak{P} \in \phi^{-1}(\{(x,y),(u,v)\})$ and $\mathfrak{P}^\circ \in \phi^{-1}(\{(x^\circ, y^\circ),(u^\circ, v^\circ)\})$. Then

$$\mathfrak{P} + \mathfrak{P}^\circ \in \phi^{-1}(\{(x,y),(u,v)\} + \{(x^\circ,y^\circ),(u^\circ,v^\circ)\}) \qquad (10.2.10)$$

and is defined over $k(\sqrt{d_1}, \sqrt{d_2}, \sqrt{d_1^\circ}, \sqrt{d_2^\circ})$. But, for each i, the action $\sqrt{d_i} \mapsto -\sqrt{d_i}$, $\sqrt{d_i^\circ} \mapsto -\sqrt{d_i^\circ}$ takes $\mathfrak{P} + \mathfrak{P}^\circ$ to $\mathbf{W}_i(\mathfrak{P}) + \mathbf{W}_i(\mathfrak{P}^\circ) = \mathfrak{P} + \mathfrak{P}^\circ$. It follows that $\mathfrak{P} + \mathfrak{P}^\circ$ is defined over $k(\sqrt{d_1 d_1^\circ}, \sqrt{d_2 d_2^\circ})$. By considering the actions of this field over k, it is easy to see that (9) is satisfied when $\mathfrak{P}$ is replaced by $\mathfrak{P} + \mathfrak{P}^\circ$ and each d_i is replaced by $d_i d_i^\circ$. We therefore have

$$\mu^\phi(\{(x,y),(u,v)\} + \{(x^\circ,y^\circ),(u^\circ,v^\circ)\}) = [d_1 d_1^\circ, d_2 d_2^\circ],$$
$$\mu^\phi(\{(x,y),(u,v)\})\mu^\phi(\{(x^\circ,y^\circ),(u^\circ,v^\circ)\}) = [d_1 d_1^\circ, d_2 d_2^\circ]. \qquad (10.2.11)$$

As with elliptic curves, the facts $\mu^\phi(-\{(x,y),(u,v)\}) = [d_1, d_2] = [d_1, d_2]^{-1}$ and $\mu^\phi(\mathfrak{O}) = 1$ are trivial; these, together with (11), give that μ^ϕ is a group homomorphism. Further, $\{(x,y),(u,v)\}$ is in the kernel of μ^ϕ precisely when $\mathfrak{P} \in \mathfrak{G}$. Therefore, the induced map

$$\mu^\phi : \quad \widehat{\mathfrak{G}}/\phi(\mathfrak{G}) \longrightarrow k^*/(k^*)^2 \times k^*/(k^*)^2,$$
$$\{(x,y),(u,v)\} \longmapsto [d_1, d_2] \qquad (10.2.12)$$

is injective. There is an alternative definition of μ^ϕ already indicated in Chapter 9, Section 9. Namely

$$\mu^\phi : \quad \widehat{\mathfrak{G}}/\phi(\mathfrak{G}) \longrightarrow k^*/(k^*)^2 \times k^*/(k^*)^2,$$
$$\{(x,y),(u,v)\} \longmapsto [L_1(x)L_1(u), L_2(x)L_2(u)], \qquad (10.2.13)$$

where L_1, L_2 are the quadratics in (1.10). In the notation of Section 1, we have $d_1 = v_4^\star/v_0^\star$, $d_2 = v_8^\star/v_0^\star$ and $d_3 = v_{12}^\star/v_0^\star$. Again, it is the second description (13) which is better than (12) for computing the value of a given $\mu^\phi(\{(x,y),(u,v)\})$. There is also the equivalent extended version

$$\mu_{\text{ext}}^\phi : \quad \widehat{\mathfrak{G}}/\phi(\mathfrak{G}) \longrightarrow k^*/(k^*)^2 \times k^*/(k^*)^2 \times k^*/(k^*)^2,$$
$$\{(x,y),(u,v)\} \longmapsto [L_1(x)L_1(u), L_2(x)L_2(u), L_3(x)L_3(u)]. \qquad (10.2.14)$$

The third component is redundant, since it is the product of the first two components in $k^*/(k^*)^2$, but it is a useful device for computing μ^ϕ at those $\{(x,y),(u,v)\} \in \widehat{\mathfrak{G}}$ for which either of $L_1(x)L_1(u)$ or $L_2(x)L_2(u)$ is zero. In future computations, we shall always write members of the image of μ^ϕ in the abbreviated form of (13). Using μ^ϕ_{ext} as a device to compute μ^ϕ, we can see that the 2-torsion members of $\widehat{\mathfrak{G}}$ corresponding to L_1, L_2, L_3 map under μ^ϕ to $[\hat{b}_{12}\hat{b}_{13}, \hat{b}_{12}]$, $[\hat{b}_{12}, \hat{b}_{12}\hat{b}_{23}]$, $[\hat{b}_{13}, \hat{b}_{23}]$ respectively. Applying the same argument to $\mu^{\hat{\phi}}$, the corresponding injection on $\mathfrak{G}/\hat{\phi}(\widehat{\mathfrak{G}})$ gives the genus 2 version of (7).

$$
\begin{aligned}
&\{[1,1], [b_{12}b_{13}, b_{12}], [b_{12}, b_{12}b_{23}], [b_{12}, b_{12}b_{23}]\} \leqslant \operatorname{im} \mu^\phi, \\
&\{[1,1], [\hat{b}_{12}\hat{b}_{13}, \hat{b}_{12}], [\hat{b}_{12}, \hat{b}_{12}\hat{b}_{23}], [\hat{b}_{12}, \hat{b}_{12}\hat{b}_{23}]\} \leqslant \operatorname{im} \hat{\mu}^\phi,
\end{aligned}
\tag{10.2.15}
$$

where the $b_{ij}, \hat{b}_{ij}$ are as in (1.11).

Example 10.2.1. Let $\mathcal{C}_2$ be the curve of (8.1.12), namely

$$
\mathcal{C}_2 : \ Y^2 = G_1(X)G_2(X)G_3(X) = (X^2+1)(X^2+2)(X^2+X+1). \tag{10.2.16}
$$

Here, in the notation of (9.2.2), $g_{10} = 1$, $g_{11} = 0$, $g_{12} = 1$, $g_{20} = 2$, $g_{21} = 0$, $g_{22} = 1$, $g_{30} = 1$, $g_{31} = 1$ and $g_{32} = 1$, so that

$$
\Delta = \det(g_{ij}) = \det\begin{pmatrix} 1 & 0 & 1 \\ 2 & 0 & 1 \\ 1 & 1 & 1 \end{pmatrix} = 1,
$$

$$
\begin{aligned}
L_1 &= G_2'(X)G_3(X) - G_2(X)G_3'(X) = X^2 - 2X - 2, \\
L_2 &= G_3'(X)G_1(X) - G_3(X)G_1'(X) = -X^2 + 1, \\
L_3 &= G_1'(X)G_2(X) - G_1(X)G_2'(X) = 2X.
\end{aligned}
\tag{10.2.17}
$$

The curve of (9.2.3) with isogenous jacobian is then

$$
\widehat{\mathcal{C}}_2 : \ \Delta Y^2 = L_1(X)L_2(X)L_3(X) = (X^2-2X-2)(-X^2+1)(2X). \tag{10.2.18}
$$

Then from (13) we can compute, for example,

$$
\begin{aligned}
\mu^\phi : \ \{(0,0), (-\tfrac{1}{2}, \tfrac{3}{4})\} &\longmapsto [L_1(0)L_1(-\tfrac{1}{2}), L_2(0)L_2(-\tfrac{1}{2})] \\
&= [\tfrac{3}{2}, \tfrac{3}{4}] \\
&= [6, 3] \in \mathbb{Q}^*/(\mathbb{Q}^*)^2 \times \mathbb{Q}^*/(\mathbb{Q}^*)^2.
\end{aligned}
\tag{10.2.19}
$$

To find the image of $\{(1,0),(-1,0)\}$, we first compute $L_1(1)L_1(-1) = -3$ and $L_3(1)L_3(-1) = -1$, from which we can deduce that the second component is $(-3)(-1)$, so that

$$\mu^\phi : \ \{(1,0),(-1,0)\} \longmapsto [-3,(-3)(-1)] = [-3,3]. \qquad (10.2.20)$$

To find the image of $\{(1,0),(0,0)\}$, this trick is not quite possible, since both the second and third components of (14) are undefined. So we find first $\mu^\phi((1,0)) = [-3,(-3)(-2)] = [-3,6]$, and then $\mu^\phi((0,0)) = [-2,-1]$. Therefore

$$\mu^\phi : \ \{(1,0),(0,0)\} \longmapsto [-3,6][-2,-1] = [6,-6] \qquad (10.2.21)$$

From (15), the images under $\mu^{\hat\phi}$ of the 3 divisors of order 2 in $\mathfrak{G}$ are as follows.

$$\mu^{\hat\phi} : \ \{(i,0),(-i,0)\} \longmapsto [1,1],$$
$$\mu^{\hat\phi} : \ \{(\sqrt{-2},0),(-\sqrt{-2},0)\} \longmapsto [1,3], \qquad (10.2.22)$$
$$\mu^{\hat\phi} : \ \{(-\tfrac{1}{2}+\tfrac{1}{2}\sqrt{-3},0),(-\tfrac{1}{2}-\tfrac{1}{2}\sqrt{-3},0)\} \longrightarrow [1,3].$$

Of course, the third of these could also have been computed as the product of the first two. Note that these maps can be used to find all 2-power torsion in $\mathfrak{G}$. Since the three quadratics in (16) are irreducible, the divisors in (22) give a complete list of rational 2-torsion elements in $\mathfrak{G}$. The second and third of these do not map to $[1,1]$ under $\mu^{\hat\phi}$, and so are not in $\hat\phi(\widehat{\mathfrak{G}})$, and are not in $2\mathfrak{G} = \hat\phi \circ \phi(\mathfrak{G})$. The first of these, namely $\{(i,0),(-i,0)\}$, does map to $[1,1]$ under $\mu^{\hat\phi}$ and so certainly is in $\hat\phi(\widehat{\mathfrak{G}})$. Indeed we can compute directly that the preimages of $\{(i,0),(-i,0)\}$ under $\hat\phi$ are precisely

$$\{(1,0),(0,0)\}, \{(1,0),\infty\}, \{(-1,0),(0,0)\}, \{(-1,0),\infty\}. \qquad (10.2.23)$$

The images of these under μ^ϕ are $[6,-6]$, $[-3,6]$, $[-2,-2]$, $[1,2]$, respectively. Therefore, none of these are in $\phi(\mathfrak{G})$, and so $\{(i,0),(-i,0)\}$ is not in $2\mathfrak{G}$. This shows that none of the 2-torsion elements of $\mathfrak{G}$ are in $2\mathfrak{G}$, and so there is no 4-torsion in $\mathfrak{G}$, nor any 2^k-torsion for $k > 1$.

This observation eases the computation of all torsion in $\mathfrak{G}$. Recall from Example 8.2.2 that reductions modulo 5 and 7 were sufficient to show that the entire torsion group of $\mathfrak{G}$ looks like $C_2 \times C_4$ if $\{(i,0),(-i,0)\}$ lies in $2\mathfrak{G}$ and is $C_2 \times C_2$ otherwise. The above comments using the isogeny show that the latter must be the case, without requiring any further finite field computations. This also explains why, in Example 8.2.2, it was necessary to go all the way up to $p = 13$, this being the smallest prime of good reduction for which none of $[6,-6], [-3,6], [-2,-2], [1,2]$ are equal to $[1,1]$ in $\mathbb{F}_p^*/(\mathbb{F}_p^*)^2 \times \mathbb{F}_p^*/(\mathbb{F}_p^*)^2$.

3   Homogeneous spaces $J^{\phi}_{d_1,d_2}$ and norm spaces $L^{\phi}_{d_1,d_2}$.

(i) Elliptic curves. Suppose that $\mathcal{E}$ and $\widehat{\mathcal{E}}$ are as in (1.1) and (1.7), and $d \in \mathrm{im}\,\mu^{\phi}$; $d = \mu^{\phi}\big((x,y)\big)$, say. Using the definition of (2.1),(2.2), there must exist $(x,y) \in \widehat{\mathfrak{G}} = \widehat{\mathcal{E}}(k)$ and $P \in \mathcal{E}\big(k(\sqrt{d})\big)$ such that $\phi(P) = (x,y)$, and $\mathbf{W}(P)$ is equal to the conjugate of P. Now let $(z_i) = (z_i(P)) \in \mathbb{P}^3\big(k(\sqrt{d})\big)$ be the vector corresponding to P using the $\mathbb{P}^3$ embedding of (7.1.2), and let $(v_i) = (v_i(P)) \in \mathbb{P}^3\big(k(\sqrt{d})\big)$ be with respect to the new coordinates given in (1.4). Then $\mathbf{W}$, and hence conjugation also, has the effect given in (1.5). Let

$$w_i = v_i, \; i = 0,1, \qquad w_i = v_i/\sqrt{d}, \; i = 2,3. \qquad (10.3.1)$$

Then conjugation leaves (w_i) unchanged, so that $(w_i) \in \mathbb{P}^3(k)$. Replacing (v_i) for (w_i) in (1.6) gives

$$J^{\phi}_d : \qquad \begin{cases} dw_2^2 = w_0^2 - 4bw_1^2 \\ dw_3^2 = w_0v_1 + aw_1^2. \end{cases} \qquad (10.3.2)$$

The above steps are all reversible, so the converse is also true; namely, if there exists a $(w_i) \in \mathbb{P}^3(k)$ satisfying (2) then $d \in \mathrm{im}\,\mu^{\phi}$. This can be summarized by the equivalence

$$d \in \mathrm{im}\,\mu^{\phi} \Leftrightarrow J^{\phi}_d(k) \neq \emptyset \qquad (10.3.3)$$

The spaces J of (1.2) and J^{ϕ}_d of (2) (a *homogeneous space*) are *twists* of each other: they are all defined over k, but the birational transformations between them is defined over $k(\sqrt{d})$.

The first of the two equations of (2) can be separated off to define the *norm space*

$$L^{\phi}_d : \quad dw_2^2 = w_0^2 - 4bw_1^2 \qquad (10.3.4)$$

Clearly $L^{\phi}_d(k)$ being non-empty is a prerequisite for $J^{\phi}_d(k)$ to be non-empty. The condition that $L^{\phi}_d(k)$ is non-empty can be expressed in terms of the following symmetric relationship between $k(\sqrt{b})$ and $k(\sqrt{d})$.

$$\begin{aligned} L^{\phi}_d(k) \neq \emptyset &\Leftrightarrow \exists \rho \in \mathbb{Q}(\sqrt{d}), \quad b = N_{\mathbb{Q}(\sqrt{d})/\mathbb{Q}}(\rho) \\ &\Leftrightarrow \exists \rho' \in \mathbb{Q}(\sqrt{b}), \quad d = N_{\mathbb{Q}(\sqrt{b})/\mathbb{Q}}(\rho'). \end{aligned} \qquad (10.3.5)$$

When $k = \mathbb{Q}$ then L^{ϕ}_d, unlike J^{ϕ}_d, satisfies the Hasse principle.¶

¶ It is an exercise for the reader [or see Flynn (1994), p. 29] to show that mere consideration of norms gives rank $\leqslant 1$ for $Y^2 = X^3 + pX$ when $p \equiv 3, 5$ mod 8 (and so one has an easy proof, for example, that $Y^2 = X^3 + 3X$ has rank 1, since it has the point $(1,2)$ of infinite order). See also [Flynn (1994), pp. 37–39] for analogous genus 2 computations.

(ii) Genus 2. Returning to our curves of genus 2, C and $\hat{C}$ of (1.10), suppose that $[d_1, d_2] \in \operatorname{im} \mu^\phi$. Then there exist $\{(x, y), (u, v)\}$ and $\mathfrak{P}$ satisfying (2.9). Letting $(z_i) = (z_i(\mathfrak{P}))$ and $(v_i) = (v_i(\mathfrak{P})) \in \mathbb{P}^{15}(k(\sqrt{d_1}, \sqrt{d_2}))$ be as described just above (1.12), then the maps $\mathbf{W}_1$, $\mathbf{W}_2$ and $\mathbf{W}_3$ are the linear maps given in (1.12). Each $\mathbf{W}_i$ has the same effect as conjugation in $k(\sqrt{d_1}, \sqrt{d_2})$ over $k(\sqrt{d_i})$ [recall that we have defined $d_3 = d_1 d_2$]. Let

$$w_i = \begin{cases} v_i, & i = 0, \dots, 3, \\ v_i/\sqrt{d_1}, & i = 4, \dots, 7, \\ v_i/\sqrt{d_2}, & i = 8, \dots, 11, \\ v_i/\sqrt{d_3}, & i = 12, \dots, 15. \end{cases} \tag{10.3.6}$$

Then (w_i) is invariant under all conjugations in $k(\sqrt{d_1}, \sqrt{d_2})$ over $k(\sqrt{d_i})$ for $i = 1, 2, 3$. Since these generate $\operatorname{Gal}(k(\sqrt{d_1}, \sqrt{d_2})/k)$, we must have $(w_i) \in \mathbb{P}^{15}(k)$. We can let $J^\phi_{d_1, d_2}$ (a *homogeneous space*) be defined by the equations in (w_i) which, like the defining equations in (z_i), are given by 72 quadratics forms over k. As in the elliptic curve case, J_{d_1, d_2} is a twist of J; both are defined over k, but the map from J to J_{d_1, d_2}, given by (6), is defined over $k(\sqrt{d_1}, \sqrt{d_2})$. There are 20 special equations which perform the role analogous to (4) and define $L^\phi_{d_1, d_2}$. This is described in detail in Flynn (1994). The criterion analogous to (5) is as follows.

$$L^\phi_{d_1, d_2}(k) \neq \emptyset$$
$$\Leftrightarrow \exists \rho_i \in \mathbb{Q}(\sqrt{d_i}), i = 1, 2, 3,$$
$$b_1 = N_{\mathbb{Q}(\sqrt{d_2})/\mathbb{Q}}(\rho_2) N_{\mathbb{Q}(\sqrt{d_3})/\mathbb{Q}}(\rho_3),$$
$$b_2 = N_{\mathbb{Q}(\sqrt{d_1})/\mathbb{Q}}(\rho_1) N_{\mathbb{Q}(\sqrt{d_3})/\mathbb{Q}}(\rho_3), \tag{10.3.7}$$
$$\Leftrightarrow \exists \rho'_i \in \mathbb{Q}(\sqrt{b_i}), i = 1, 2, 3,$$
$$d_1 = N_{\mathbb{Q}(\sqrt{b_2})/\mathbb{Q}}(\rho'_2) N_{\mathbb{Q}(\sqrt{b_3})/\mathbb{Q}}(\rho'_3),$$
$$d_2 = N_{\mathbb{Q}(\sqrt{b_1})/\mathbb{Q}}(\rho'_1) N_{\mathbb{Q}(\sqrt{b_3})/\mathbb{Q}}(\rho'_3),$$

where as usual the b_i are as defined in (1.11). We shall not make any further use of homogeneous spaces here, since our worked examples in the next chapter will use a method independent of them.

4   Computing $\# \hat{\mathfrak{G}}/\phi(\mathfrak{G}) \cdot \# \mathfrak{G}/\hat{\phi}(\hat{\mathfrak{G}})$ when $k = \mathbb{Q}_p$. Let J represent either the elliptic curve $\mathcal{E}$ of (1.1) or the jacobian of C of (1.10); similarly for $\hat{J}$. For this section we shall assume that the ground field is $\mathbb{Q}_p$, so that $\mathfrak{G} = J(\mathbb{Q}_p)$ and $\hat{\mathfrak{G}} = \hat{J}(\mathbb{Q}_p)$. In either the genus 1 or 2 situation, there is the sequence

$$\mathfrak{G} \xrightarrow{\phi} \widehat{\mathfrak{G}} \xrightarrow{\hat{\phi}} \mathfrak{G}, \qquad [2] = \hat{\phi} \circ \phi, \tag{10.4.1}$$

on $\mathfrak{G}$, where $[2]$ represents the multiplication-by-2 map From (7.5.8) when $p \neq \infty$ and (7.6.2) when $p = \infty$, we know that $\mathfrak{G}/2\mathfrak{G}$ and $\widehat{\mathfrak{G}}/2\widehat{\mathfrak{G}}$ are finite, and so $\mathfrak{G}/\hat{\phi}(\widehat{\mathfrak{G}})$ and $\widehat{\mathfrak{G}}/\phi(\mathfrak{G})$ must both be finite also. The map between finite groups induced by $\hat{\phi}$

$$\widehat{\mathfrak{G}}/\phi(\mathfrak{G}) \longrightarrow \hat{\phi}(\widehat{\mathfrak{G}})/2\mathfrak{G} \tag{10.4.2}$$

is surjective and has kernel of order $\#\ker \hat{\phi}/(\ker \hat{\phi} \cap \phi(\mathfrak{G}))$, giving¶

$$\#\widehat{\mathfrak{G}}/\phi(\mathfrak{G}) = \#\hat{\phi}(\widehat{\mathfrak{G}})/2\mathfrak{G} \cdot \#\ker \hat{\phi}/(\ker \hat{\phi} \cap \phi(\mathfrak{G})). \tag{10.4.3}$$

The map between finite groups, restricting ϕ to $\mathfrak{G}[2]$,

$$\mathfrak{G}[2] \longrightarrow \ker \hat{\phi} \cap \phi(\mathfrak{G}), \tag{10.4.4}$$

where $\mathfrak{G}[2]$ denotes the 2-torsion group of $\mathfrak{G}$, is surjective and has kernel of order $\#\ker \phi \cap \mathfrak{G}[2] = \#\ker \phi$ (since $\ker \phi \subset \mathfrak{G}[2]$). This gives

$$\#\mathfrak{G}[2] = \#\ker \hat{\phi} \cap \phi(\mathfrak{G}) \cdot \#\ker \phi. \tag{10.4.5}$$

Combining (4) and (5) with

$$\#\mathfrak{G}/\hat{\phi}(\widehat{\mathfrak{G}}) \cdot \#\hat{\phi}(\widehat{\mathfrak{G}})/2\mathfrak{G} = \#\mathfrak{G}/2\mathfrak{G} \tag{10.4.6}$$

gives the identity

$$\#\widehat{\mathfrak{G}}/\phi(\mathfrak{G}) \cdot \#\mathfrak{G}/\hat{\phi}(\widehat{\mathfrak{G}}) \cdot \#\mathfrak{G}[2] = \#\ker \phi \cdot \#\ker \hat{\phi} \cdot \#\mathfrak{G}/2\mathfrak{G}. \tag{10.4.7}$$

Combining this with the Corollary to (7.5.8) and (7.6.2), and the fact that $\#\ker \phi = \#\ker \hat{\phi} = 2^g$, gives

$$\#\widehat{\mathfrak{G}}/\phi(\mathfrak{G}) \cdot \#\mathfrak{G}/\hat{\phi}(\widehat{\mathfrak{G}}) = \begin{cases} 2^g & \text{for } p = \infty; \\ 4^g & \text{for } p \neq 2, \infty; \\ 8^g & \text{for } p = 2, \end{cases} \tag{10.4.8}$$

or, more succinctly,

$$\#\widehat{\mathfrak{G}}/\phi(\mathfrak{G}) \cdot \#\mathfrak{G}/\hat{\phi}(\widehat{\mathfrak{G}}) = \left(4/|2|_p\right)^g \tag{10.4.9}$$

¶ See footnote on page 72.

for any p. This gives an easy way to recognize when a complete set of generators for both $\widehat{\mathfrak{G}}/\phi(\mathfrak{G})$ and $\mathfrak{G}/\hat{\phi}(\widehat{\mathfrak{G}})$ have been found.

In a similar manner to (7.5.9), the above equations $(2),\ldots,(7)$ can be summarized by the following Snake Lemma diagram:

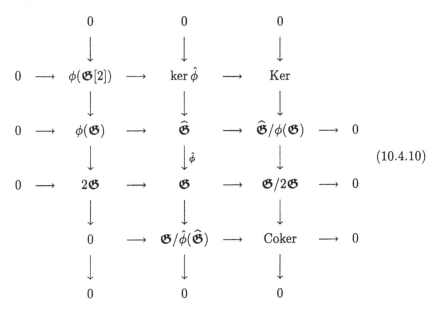

$$(10.4.10)$$

where the maps corresponding to the middle row of down arrows are all induced by $\hat{\phi}$. The entries Ker and Coker should be defined in the unique way which makes their column exact; that is, they are defined to be

$$\ker\big(\widehat{\mathfrak{G}}/\phi(\mathfrak{G})\xrightarrow{\hat{\phi}}\mathfrak{G}/2\mathfrak{G}\big),\quad (\mathfrak{G}/2\mathfrak{G})/\mathrm{im}\big(\widehat{\mathfrak{G}}/\phi(\mathfrak{G})\xrightarrow{\hat{\phi}}\mathfrak{G}/2\mathfrak{G}\big),\quad (10.4.11)$$

respectively. The Snake Lemma says that we can join the top and bottom nontrivial rows into an exact sequence. Clearly, from this sequence, we have that $\#\mathrm{Ker}=\#\ker\hat{\phi}/\#\phi(\mathfrak{G}[2])$ and $\#\mathrm{Coker}=\#\mathfrak{G}/\hat{\phi}(\widehat{\mathfrak{G}})$. From this, together with the sequence given by the right hand nontrivial column, and the straightforward substitution of $\#\mathfrak{G}[2]/\#\ker\phi$ in place of $\#\phi(\mathfrak{G}[2])$, we can again obtain (7).

Chapter 11

Computing the Mordell-Weil group

0   Introduction. We shall sketch a proof of the finiteness of $\mathfrak{G}/2\mathfrak{G}$ when the ground field is a number field. We shall outline how, in many cases, a descent can be performed to compute $\mathfrak{G}/2\mathfrak{G}$, and hence the rank of $\mathfrak{G}$, with a quick 'pen and paper' computation, not requiring the use of homogeneous spaces. This is illustrated at the end of the chapter with a worked example of a complete 2-descent over $\mathbb{Q}$ and a descent via isogeny over $\mathbb{Q}$.

1   The Weak Mordell-Weil Theorem. As in Chapter 10, let

$$
\begin{aligned}
\mathcal{E} &: \quad Y^2 = X(X^2 + aX + b), \\
\widehat{\mathcal{E}} &: \quad Y^2 = X(X^2 + \hat{a}X + \hat{b}),
\end{aligned}
\tag{11.1.1}
$$

where $\hat{a} = -2a, \hat{b} = a^2 - 4b$, and

$$
\begin{aligned}
\mathcal{C} &: \quad Y^2 = F(X) = G_1(X)G_2(X)G_3(X), \\
\widehat{\mathcal{C}} &: \quad \Delta Y^2 = H(X) = L_1(X)L_2(X)L_3(X),
\end{aligned}
\tag{11.1.2}
$$

where the G_i, L_i, Δ are as in Chapter 9, Section 2. As usual, let $\mathfrak{G}$ denote the group of rational points on the jacobian over the ground field. For the moment we shall assume that all of the above curves are defined over $\mathbb{Q}$, so that we can take all of the coefficients to be in $\mathbb{Z}$, and $\mathfrak{G} = J(\mathbb{Q})$, $\widehat{\mathfrak{G}} = \widehat{J}(\mathbb{Q})$. In the elliptic curve case, let

$$
S = \{p : p \mid b(a^2 - 4b)\} \cup \{2\} = \{p_1, \ldots, p_r\},
\tag{11.1.3}
$$

and in the genus 2 case let

$$
S = \{p : p \mid \Delta b_1 b_2 b_3 \hat{b}_1 \hat{b}_2 \hat{b}_3\} \cup \{2\} = \{p_1, \ldots, p_r\},
\tag{11.1.4}
$$

where the $b_i, \hat{b}_i$ are as in (10.1.11). This includes all primes of bad reduction of the above curves. In either case, let

$$
\mathbb{Q}(S) = \{\pm p_1^{e_1} \ldots p_r^{e_r} : e_i = 0, 1\} \leqslant \mathbb{Q}^*/(\mathbb{Q}^*)^2.
\tag{11.1.5}
$$

Let d (in the genus 1 case) or $[d_1, d_2]$ (in the genus 2 case) be in the image of μ^ϕ; the image of $\mathfrak{R} \in \widehat{\mathfrak{G}}$, say. Represent each of $d, d_1, d_2 \in \mathbb{Q}^*/(\mathbb{Q}^*)^2$ by square free integers. Let

$$L = \mathbb{Q}(\sqrt{d}), \text{ when } g = 1, \quad L = \mathbb{Q}(\sqrt{d_1}, \sqrt{d_2}) \text{ when } g = 2. \quad (11.1.6)$$

From the definition of μ^ϕ, given in (10.2.1), (10.2.2) for $g = 1$, and (10.2.8), (10.2.9) for $g = 2$, there must exist $\mathfrak{P} \in J(L)$ and $\mathbf{W}_i$ such that $\phi(\mathfrak{P}) = \mathfrak{R}$ and

$$\mathbf{W}_i(\mathfrak{P}) = \sigma(\mathfrak{P}), \quad (11.1.7)$$

where $\mathbf{W}_i$ represents addition by a point of order 2, and σ represents, in the genus 1 case, the conjugation map $\sqrt{d} \mapsto -\sqrt{d}$ on L, and in the genus 2 case, the conjugation map defined by $\sqrt{d_1} \mapsto -\sqrt{d_1}$, $\sqrt{d_2} \mapsto \sqrt{d_2}$ on L. Write

$$\mathbf{z} = (z_i) = (z_i(\mathfrak{P})) \in \mathbb{P}^{4^g - 1}(L), \quad (11.1.8)$$

where, as usual, the z_i are given by the $z_0, \ldots, z_3$ of (7.1.2) in the elliptic curve case, and by the coordinates $z_0, \ldots, z_{15}$ described at the beginning of Chapter 2, Section 2 in the genus 2 case. Suppose that there is a prime p not in $\mathbb{Q}(\mathcal{S})$ dividing d, when $g = 1$, or dividing either of d_1, d_2 (without loss of generality, say d_1) when $g = 2$. We can consider (z_i) to be a member of $\mathbb{P}^{4^g - 1}(L_\mathfrak{p})$, where $\mathfrak{p}$ is a prime above p. Write (z_i) so that $\max |z_i|_\mathfrak{p} = 1$ and let $(\tilde{z}_i)$ be the reduction of (z_i) mod $\mathfrak{p}$. Since $|\sqrt{d_1}|_\mathfrak{p} < 1$ it follows that $\sigma(\tilde{z}_i) = (\tilde{z}_i)$, so that

$$\widetilde{\mathbf{W}}_i(\tilde{\mathbf{z}}) = \tilde{\mathbf{z}} \quad (11.1.9)$$

on $\tilde{J}$, the reduction of J mod $\mathfrak{p}$. But this implies that one of the points of order 2 on $\tilde{J}$ is the same as $\tilde{\mathfrak{O}}$, which is only possible if two roots of the reduced curve mod $\mathfrak{p}$ are equal, contradicting the assumption that p is a prime of good reduction.

This contradiction shows that, in genus $g = 1, 2$:

$$\operatorname{im} \mu^\phi \leqslant (\mathbb{Q}(\mathcal{S}))^{\times g}. \quad (11.1.10)$$

where, for any group G and integer ℓ, the notation $G^{\times \ell}$ always means

$$G^{\times \ell} = \underbrace{G \times \ldots \times G}_{\ell \text{ times}} \quad (11.1.11)$$

From (10) we immediately have the finiteness of the image of the induced injection μ^ϕ on $\widehat{\mathfrak{G}}/\phi(\mathfrak{G})$, given in (10.2.4), (10.2.5) for elliptic curves and (10.2.12), (10.2.13) in genus 2. Therefore $\widehat{\mathfrak{G}}/\phi(\mathfrak{G})$ is finite. Similarly, $\mu^{\hat{\phi}}$

is contained in the same finite set as given in (10), giving that $\mathfrak{G}/\hat{\phi}(\widehat{\mathfrak{G}})$ is finite also. Since $[2] = \hat{\phi} \circ \phi$, it follows that $\mathfrak{G}/2\mathfrak{G}$ is finite.

Alternatively, suppose that $Y^2 = F(X)$, our elliptic curve or curve of genus 2, is such that all roots of $F(X)$ are rational. Then the curve can be written in the form

$$\mathcal{C}: \quad Y^2 = \prod_{i=1}^{2g+1} (X - e_i), \quad e_i \in \mathbb{Z}. \tag{11.1.12}$$

We would also like an explicit description of the image in this situation, since two of the worked examples (Examples 2.1, 3.1) will be of this form. Using Chapter 6, Section 10, the usual injection μ' on $\mathfrak{G}/2\mathfrak{G}$, where $\mathfrak{G} = J(\mathbb{Q})$, can be taken to be

$$\mu': \ \mathfrak{G}/2\mathfrak{G} \longrightarrow \left(\mathbb{Q}^*/(\mathbb{Q}^*)^2\right)^{\times 2g},$$
$$(x,y) - \infty \longmapsto [x - e_1, \ldots, x - e_{2g}], \tag{11.1.13}$$

Of course, it is equivalent to use the map

$$\mu'_{\text{ext}}: \ \mathfrak{G}/2\mathfrak{G} \longrightarrow \left(\mathbb{Q}^*/(\mathbb{Q}^*)^2\right)^{\times(2g+1)},$$
$$(x,y) - \infty \longmapsto [x - e_1, \ldots, x - e_{2g}, x - e_{2g+1}], \tag{11.1.14}$$

where the final component is redundant, being equal to the product of the other components. In the genus 2 case, the map on $\{(x,y),(u,v)\} = (x,y) + (u,v) - 2\infty$ is just defined to be $\mu\big((x,y) - \infty\big)\mu\big((u,v) - \infty\big)$. In this case, take

$$\mathcal{S} = \{p : p \mid \prod_{i \neq j}(e_i - e_j)\} \cup \{2\} = \{p_1, \ldots, p_r\},$$
$$\mathbb{Q}(\mathcal{S}) = \{\pm p_1^{e_1} \ldots p_r^{e_r} : e_i = 0, 1\} \leqslant \mathbb{Q}^*/(\mathbb{Q}^*)^2, \tag{11.1.15}$$

in which case we have the similar result that

$$\operatorname{im}\mu' \leqslant \left(\mathbb{Q}(\mathcal{S})\right)^{\times 2g}. \tag{11.1.16}$$

The generalization to the case where the curve is defined over a number field K is essentially the same as the standard argument for elliptic curves, such as that on p. 70 of Cassels (1991), for example. One can adjoin the roots of $F(X)$ to K to give a number field over which $F(X)$ can be written in the form (12), where the e_i are algebraic integers. One can then imitate the above style of argument on ideals, and then use the finiteness of the class number to show that $\operatorname{im}\mu'$ is again finite.

In the case when $F(X)$ is a quintic over a number field K, with factorization over K given by $F(X) = F_1(X) \cdot \ldots \cdot F_s(X)$, take S to be the finite set of primes of K that includes the infinite primes, the primes lying over 2 and the primes of bad reduction for J. Let L_i denote K with a root of $F_i(X)$ adjoined. Then the map μ' of Theorem 6.10.1 is into $K(\Theta)^*/(K(\Theta)^*)^2 = L^*/(L^*)^2 = L_1^*/(L_1^*)^2 \times \ldots \times L_s^*/(L_s^*)^2$, where $K(\Theta) = K[T]/F[T]$. One can show that the image of μ' is a subset of the group generated by the elements in the kernel of the norm from $K(\Theta)^*/(K(\Theta)^*)^2$ to $K^*/(K^*)^2$ which have the property that when you adjoin the square roots of their images in the fields L_i to the fields L_i you get extensions that are ramified only at the primes of S. For the sextic case, when K has class number 1, it has been shown in Flynn, Poonen & Schaefer (1995) that precisely the same statement remains true about the image of μ in $K(\Theta)^*/K^*(K(\Theta)^*)^2$

2   Performing 2-descent without using homogeneous spaces.

Assume, for simplicity, that the ground field is $\mathbb{Q}$. Recall that, as always, $\mathfrak{G} = J(k)$, where k is the ground field; so here, $\mathfrak{G} = J(\mathbb{Q})$. For any p, we shall also want to refer to $J(\mathbb{Q}_p)$, which we shall denote by $\mathfrak{G}_p$.¶ We begin with the standard commutative diagram

$$
\begin{array}{ccc}
\mathfrak{G}/2\mathfrak{G} & \xrightarrow{\;\mu'\;} & M \\
\Big\downarrow{\scriptstyle i_p} & & \Big\downarrow{\scriptstyle j_p} \\
\mathfrak{G}_p/2\mathfrak{G}_p & \xrightarrow{\;\mu'_p\;} & M_p
\end{array}
\qquad (11.2.1)
$$

where the bottom row is constructed in the same way as the top row, but with respect to $\mathbb{Q}_p$. The maps i_p and j_p are natural maps on the quotient induced by the inclusion map from $\mathbb{Q}$ into $\mathbb{Q}_p$ (note that i_p and j_p are not injective in general). The set M is a finite set which is known to contain the image of μ', as discussed in the last section. For example, when $F(X)$ has all of its roots in $\mathbb{Q}$, then we can take M to be the finite subset of $(\mathbb{Q}^*/(\mathbb{Q}^*)^2)^{\times 2g}$ given in (1.15), and M_p to be $(\mathbb{Q}_p^*/(\mathbb{Q}_p^*)^2)^{\times 2g}$. Suppose that one has, after a short search, found some members of $\mathfrak{G}$, and suspects that these generate $\mathfrak{G}/2\mathfrak{G}$. In practice, a search in $\mathfrak{G}_p$ quickly finds a set

---

¶ We draw the reader's attention to the fact that in Chapter 7, Sections 4, 5, 6 and Chapter 10, Section 4, the ground field was a finite extension of some $\mathbb{Q}_p$, and so the results about the groups denoted $\mathfrak{G}$ in those sections, will now apply to the groups denoted $\mathfrak{G}_p$ in this section.

of generators for $\mathfrak{G}_p/2\mathfrak{G}_p$. Recall that (7.6.2) and the corollary to (7.5.8) give¶

$$\#\mathfrak{G}_p/2\mathfrak{G}_p = \begin{cases} \#\mathfrak{G}_p[2]/2^g, & p = \infty; \\ \#\mathfrak{G}_p[2], & p \neq 2, \infty; \\ \#\mathfrak{G}_p[2] \cdot 2^g, & p = 2. \end{cases} \qquad (11.2.2)$$

This can be summarized as

$$\#\mathfrak{G}_p/2\mathfrak{G}_p = \#\mathfrak{G}_p[2]/|2|_p^g, \text{ for all } p, \qquad (11.2.3)$$

which tells us when we have found all of $\mathfrak{G}_p/2\mathfrak{G}_p$. We can then compute $j_p^{-1}(\text{im}\,\mu_p')$, which must contain im μ'. If, after repeating this process for a selection of primes, the intersection of the resulting $j_p^{-1}(\text{im}\,\mu_p')$ with M is the same as the image of the known members of $\mathfrak{G}/2\mathfrak{G}$, then $\mathfrak{G}/2\mathfrak{G}$ will be completely determined. Of course, for either elliptic curves or jacobians of curves of genus 2, the group $\mathfrak{G}$ is finitely generated, as we shall see in Chapter 12. The torsion subgroup, $\mathfrak{G}_{\text{tors}}$, of $\mathfrak{G}$ is finite, by Lemma 8.2.1. So we can write

$$\mathfrak{G} \cong \mathfrak{G}_{\text{tors}} \times \mathbb{Z}^{\times r},$$
$$\mathfrak{G}/2\mathfrak{G} \cong (\mathfrak{G}_{\text{tors}}/2\mathfrak{G}_{\text{tors}}) \times C_2^{\times r}, \qquad (11.2.4)$$

where r is the *rank* of $\mathfrak{G}$ and $C_2 = \mathbb{Z}/2\mathbb{Z}$. Clearly the determination of $\mathfrak{G}_{\text{tors}}$ and $\mathfrak{G}/2\mathfrak{G}$ is sufficient to find the rank of $\mathfrak{G}$, although it does not in general give us a set of generators for $\mathfrak{G}$. The problem of deducing actual generators for $\mathfrak{G}$ (given generators for $\mathfrak{G}/2\mathfrak{G}$) will be discussed in Chapter 12. In this chapter, we concern ourselves only with finding $\mathfrak{G}_{\text{tors}}$ and $\mathfrak{G}/2\mathfrak{G}$, and hence the rank of $\mathfrak{G}$.

Before moving on to genus 2, we shall first give an elliptic curve example.

Example 11.2.1. *Let $\mathcal{E}$ be the curve $Y^2 = X(X-1)(X-4)$. Then $\mathfrak{G} = \mathcal{E}(\mathbb{Q})$ has rank 0 over $\mathbb{Q}$.*

Proof. First, $\#\widetilde{\mathcal{E}}(\mathbb{F}_5) = 4$, and so $\mathcal{E}_{\text{tors}}(\mathbb{Q}) = \{\mathfrak{O}, (0,0), (1,0), (4,0)\} \cong C_2^{\times 2}$. It is sufficient to show that these generate $\mathfrak{G}/2\mathfrak{G}$. The map μ' of (1.13) is

$$\mu' : \mathfrak{G}/2\mathfrak{G} \longrightarrow (\mathbb{Q}^*/(\mathbb{Q}^*)^2)^{\times 2}, \quad (x,y) \longmapsto [x, x-1], \qquad (11.2.5)$$

where we remind the reader that the shorthand notation $(\mathbb{Q}^*/(\mathbb{Q}^*)^2)^{\times 2}$ is as described in (1.11). The independent points $(0,0), (1,0)$ map as follows:

$$\begin{aligned} (0,0) &\longmapsto [(-1)(-4), -1] = [1, -1], \\ (1,0) &\longmapsto [1, \ (1)(-3)] = [1, -3], \end{aligned} \qquad (11.2.6)$$

¶ See footnote on page 72.

so that¶

$$H = \langle [1, -1], [1, -3] \rangle \leqslant \operatorname{im} \mu'. \tag{11.2.7}$$

The primes dividing the discriminant of $\mathcal{E}$ are 2 and 3. By (1.16), the image of μ' is contained in $\left(\mathbb{Q}(\mathcal{S})\right)^{\times 2}$, that is

$$\begin{aligned} \operatorname{im} \mu' \leqslant M &= \{\pm 1, \pm 2, \pm 3, \pm 6\}^{\times 2} \\ &= \langle [-1, 1], [2, 1], [3, 1], [1, -1], [1, 2], [1, 3] \rangle. \end{aligned} \tag{11.2.8}$$

We want to show that $H = \operatorname{im} \mu'$, which entails showing that none of the members of $M \backslash H$ can be in $\operatorname{im} \mu'$. The key point we wish to make is that there are two possible approaches here. One possibility is, for each member $[d_1, d_2]$ of $M \backslash H$ (in fact, one only has to check one representative of each coset of H in M), to look at the associated homogeneous space, as described on p. 281 of Silverman (1986), and show that it has no points in some $\mathbb{Q}_p$. We shall adopt a different point of view, and instead use (2) for each p to tell us when all of $\mathfrak{G}_p / 2 \mathfrak{G}_p$ has been found. Note that, since all possible 2-torsion occurs in $\mathfrak{G}$, we have without any further computation that, for any p,

$$\# \mathfrak{G}_p[2] = \# \mathfrak{G}[2] = 4. \tag{11.2.9}$$

First consider $p = \infty$, so that $\mathfrak{G}_p = \mathfrak{G}_\infty = \mathcal{E}(\mathbb{R})$. A set of representatives for $\mathbb{R}^* / (\mathbb{R}^*)^2$ is $\{\pm 1\}$. In diagram (1), we can take

$$\begin{aligned} M_\infty &= \left(\mathbb{R}^* / (\mathbb{R}^*)^2\right)^{\times 2} = \{\pm 1\}^{\times 2} = \langle [-1, 1], [1, -1] \rangle, \\ \ker j_\infty &= M \cap \left((\mathbb{R}^*)^2 / (\mathbb{Q}^*)^2\right)^{\times 2} = \langle [2, 1], [3, 1], [1, 2], [1, 3] \rangle. \end{aligned} \tag{11.2.10}$$

Now, the points $(0, 0), (1, 0)$, viewed as members of $\mathfrak{G}_\infty$, map under μ'_∞ to H of (7), viewed as a subgroup of M_∞. Here, $[1, -1] = [1, -3]$ but $[1, -1] \neq [1, 1]$ in $\left(\mathbb{R}^* / (\mathbb{R}^*)^2\right)^{\times 2}$. From (2) and (9), it follows that

$$\# \mathfrak{G}_\infty / 2 \mathfrak{G}_\infty = \# \mathfrak{G}_\infty[2] / 2 = 4/2 = 2, \tag{11.2.11}$$

and so the known members of $\mathfrak{G} / 2 \mathfrak{G}$ (indeed only the point $(0, 0)$ is required) are already sufficient to generate $\mathfrak{G}_\infty / 2 \mathfrak{G}_\infty$.

$$\begin{aligned} \mathfrak{G}_\infty / 2 \mathfrak{G}_\infty &= \langle (0, 0) \rangle, \\ \mu'_\infty \left(\mathfrak{G}_\infty / 2 \mathfrak{G}_\infty\right) &= \langle [1, -1] \rangle \leqslant \left(\mathbb{R}^* / (\mathbb{R}^*)^2\right)^{\times 2}. \end{aligned} \tag{11.2.12}$$

¶ We shall make free use of the standard notation $\langle g_1, \ldots, g_\ell \rangle$ to denote the subgroup generated by $g_1, \ldots, g_\ell$. Note that, since we are dealing with Boolean groups in these computations, such a subgroup will have order 2^k for some $k \leqslant \ell$.

It follows from the commutativity of (1) that

$$\begin{aligned}
\operatorname{im} \mu' &\leqslant j_\infty^{-1}\big(\mu'_\infty(\mathfrak{G}_\infty/2\mathfrak{G}_\infty)\big) = \langle \ker j_\infty, H\rangle \\
&= \langle [2,1], [3,1], [1,2], [1,3], [1,-1]\rangle.
\end{aligned}$$

(11.2.13)

Note that this allows us to exclude half of M from consideration as possible members of $\operatorname{im} \mu'$, namely those $[d_1, d_2]$ for which d_1 is negative.

Now consider $p = 3$, so that $\mathfrak{G}_p = \mathfrak{G}_3 = \mathcal{E}(\mathbb{Q}_3)$. A set of representatives for $\mathbb{Q}_3^*/(\mathbb{Q}_3^*)^2$ is $\{\pm 1, \pm 3\}$. In diagram (1), we can take

$$\begin{aligned}
M_3 &= \big(\mathbb{Q}_3^*/(\mathbb{Q}_3^*)^2\big)^{\times 2} = \{\pm 1, \pm 3\}^{\times 2} \\
&= \langle [1,-1], [1,3], [-1,1], [3,1]\rangle, \\
\ker j_3 &= M \cap \big((\mathbb{Q}_3^*)^2/(\mathbb{Q}^*)^2\big)^{\times 2} = \langle [-2,1], [1,-2]\rangle.
\end{aligned}$$

(11.2.14)

Here, $[1,-1] \neq [1,-3]$ and $[1,-1] \neq [1,1]$ in $\big(\mathbb{Q}_3^*/(\mathbb{Q}_3^*)^2\big)^{\times 2}$. From (2) and (9), it follows that

$$\#\mathfrak{G}_3/2\mathfrak{G}_3 = \#\mathfrak{G}_3[2] = 4,$$

(11.2.15)

and so the known members of $\mathfrak{G}/2\mathfrak{G}$ given in (7) are sufficient to generate $\mathfrak{G}_3/2\mathfrak{G}_3$.

$$\begin{aligned}
\mathfrak{G}_3/2\mathfrak{G}_3 &= \langle (0,0), (1,0)\rangle, \\
\mu'_3(\mathfrak{G}_3/2\mathfrak{G}_3) &= \langle [1,-1], [1,-3]\rangle \leqslant \big(\mathbb{Q}_3^*/(\mathbb{Q}_3^*)^2\big)^{\times 2}.
\end{aligned}$$

(11.2.16)

Diagram (1) gives

$$\begin{aligned}
\operatorname{im} \mu' &\leqslant j_3^{-1}\big(\mu'_3(\mathfrak{G}_3/2\mathfrak{G}_3)\big) = \langle \ker j_3, H\rangle \\
&= \langle [-2,1], [1,-2], [1,-1], [1,-3]\rangle.
\end{aligned}$$

(11.2.17)

Finally, consider $p = 2$. A set of representatives for $\mathbb{Q}_2^*/(\mathbb{Q}_2^*)^2$ is given by $\{\pm 1, \pm 2, \pm 3, \pm 6\}$. Therefore M_2 is isomorphic to M and j_2 is injective. Here, $[1,-1] \neq [1,-3]$ and $[1,-1] \neq [1,1]$ in $\big(\mathbb{Q}_2^*/(\mathbb{Q}_2^*)^2\big)^{\times 2}$. But from (2) and (9), we know that

$$\#\mathfrak{G}_2/2\mathfrak{G}_2 = \#\mathfrak{G}_2[2] \cdot 2 = 4 \cdot 2 = 8,$$

(11.2.18)

and so we are missing precisely one further generator for $\mathfrak{G}_2/2\mathfrak{G}_2$. A quick search reveals the point

$$(21, \gamma) \in \mathfrak{G}_2, \quad \gamma \in \mathbb{Q}_2^*, \quad \gamma^2 = 21 \cdot 20 \cdot 17,$$

(11.2.19)

where the existence of γ is guaranteed by the fact that $21 \cdot 5 \cdot 17 \equiv 1 \pmod 8$. Applying μ_2' gives

$$\mu_2' : (21, \gamma) \longmapsto [21, 20] = [-3, -3] \in \left(\mathbb{Q}_2^* / (\mathbb{Q}_2^*)^2 \right)^{\times 2}. \tag{11.2.20}$$

Since $[1, -1], [1, -3], [-3, -3]$ are independent members of $\left(\mathbb{Q}_2^* / (\mathbb{Q}_2^*)^2 \right)^{\times 2}$, it follows from (18) that

$$\mathfrak{G}_2 / 2\mathfrak{G}_2 = \langle (0,0), (1,0), (21, \gamma) \rangle,$$
$$\mu_2' \left(\mathfrak{G}_2 / 2\mathfrak{G}_2 \right) = \langle [1,-1], [1,-3], [-3,-3] \rangle \leqslant \left(\mathbb{Q}_2^* / (\mathbb{Q}_2^*)^2 \right)^{\times 2}, \tag{11.2.21}$$

and so, since j_2 is injective,

$$\begin{aligned} \operatorname{im} \mu' &\leqslant j_2^{-1} \left(\mu_2'(\mathfrak{G}_2 / 2\mathfrak{G}_2) \right) = \langle \ker j_2, H, [-3,-3] \rangle \\ &= \langle [1,-1], [1,-3], [-3,-3] \rangle. \end{aligned} \tag{11.2.22}$$

It is now apparent that we only require the $\mathbb{Q}_2$ information (22), together with that either at $\mathbb{R}$ (13) or at $\mathbb{Q}_3$ (17), to deduce that $H = M$, and so $\mathfrak{G}/2\mathfrak{G}$ is generated by $(0,0)$ and $(1,0)$ as required.

Of course, homogeneous spaces are useful for searching for generators. However, in examples such as the above, the entire rank computation can be done without them. This makes a big difference in genus 2, where computer algebra is required to process the homogeneous spaces, and so an imitation of the above argument (when possible) gives a much more accessible style of computation.

For descent via isogeny over $\mathbb{Q}$, the same idea applies. The relevant commutative diagrams are

$$\begin{array}{ccc} \widehat{\mathfrak{G}}/\phi(\mathfrak{G}) & \xrightarrow{\;\mu^\phi\;} & M \\[4pt] \Big\downarrow{i_p} & & \Big\downarrow{j_p} \\[4pt] \widehat{\mathfrak{G}}_p/\phi(\mathfrak{G}_p) & \xrightarrow{\;\mu_p^\phi\;} & M_p \end{array} \tag{11.2.23}$$

where M is $\left(\mathbb{Q}(S) \right)^{\times g}$ as given in (1.10), and the same diagram, but with $\widehat{\mathfrak{G}}/\phi(\mathfrak{G})$, μ^ϕ, $\widehat{\mathfrak{G}}_p/\phi(\mathfrak{G}_p)$ and μ_p^ϕ replaced by $\mathfrak{G}/\hat\phi(\widehat{\mathfrak{G}})$, $\mu^{\hat\phi}$, $\mathfrak{G}_p/\hat\phi_p(\widehat{\mathfrak{G}}_p)$ and $\mu_p^{\hat\phi}$, respectively. For each p, one searches for generators for both $\widehat{\mathfrak{G}}_p/\phi(\mathfrak{G}_p)$ and $\mathfrak{G}/\hat\phi(\widehat{\mathfrak{G}}_p)$, using (10.4.8) to decide when they have all been found. After several choices of primes, one then hopes to determine both $\widehat{\mathfrak{G}}/\phi(\mathfrak{G})$ and $\mathfrak{G}/\hat\phi(\widehat{\mathfrak{G}})$, and hence $\mathfrak{G}/2\mathfrak{G}$.

3   A worked example of complete 2-descent. In this section we shall rework an example of a complete 2-descent on the jacobian of a curve of genus 2. This example was originally due to Gordon and Grant (1993), who used computer algebra to derive explicit equations for the homogeneous spaces which arise in a complete 2-descent. However, we shall instead imitate the working used in Example 2.1, avoiding the need for homogeneous spaces. The following style of argument was first applied in genus > 1 by Schaefer (1995).

Example 11.3.1. *Let C_1 be the curve of (8.1.1); that is*

$$C_1 : \quad Y^2 = F_1(X) = X(X-1)(X-2)(X-5)(X-6). \qquad (11.3.1)$$

Then $\mathfrak{G}$ has rank 1, with $\mathfrak{G}/2\mathfrak{G} = \langle \mathfrak{G}_{\text{tors}}, \{(3,6),\infty\}\rangle$.

Proof. It was shown in Example 8.2.1 that $\mathfrak{G}_{\text{tors}}$ consists only of the 2-torsion group of order 16. We can take as four independent generators $\{(0,0),\infty\}, \{(1,0),\infty\}, \{(2,0),\infty\}$ and $\{(5,0),\infty\}$. There is also the divisor $\{(3,6),\infty\} \in \mathfrak{G}$ of infinite order. It is sufficient to show that these generate $\mathfrak{G}/2\mathfrak{G}$. The map μ' of (1.13) is

$$\mu' : \mathfrak{G}/2\mathfrak{G} \longrightarrow \left(\mathbb{Q}^*/(\mathbb{Q}^*)^2\right)^{\times 4}, (x,y) \longmapsto [x, x-1, x-2, x-5]. \quad (11.3.2)$$

The above members of $\mathfrak{G}$ map as follows.

(i) $\{(0,0),\infty\} \longmapsto [(-1)(-2)(-5)(-6), -1, -2, -5] = [15, -1, -2, -5]$,

(ii) $\{(1,0),\infty\} \longmapsto [1, (1)(-1)(-4)(-5), -1, -4] = [1, -5, -1, -1]$,

(iii) $\{(2,0),\infty\} \longmapsto [2, 1, (2)(1)(-3)(-4), -3] = [2, 1, 6, -3]$,

(iv) $\{(5,0),\infty\} \longmapsto [5, 4, 3, (5)(4)(3)(-1)] = [5, 1, 3, -15]$,

(v) $\{(3,6),\infty\} \longmapsto [3, 2, 1, -2]$, $(11.3.3)$

so that

$$\begin{aligned} H = \langle &[15, -1, -2, -5], [1, -5, -1, -1], [2, 1, 6, -3], \\ &[5, 1, 3, -15], [3, 2, 1, -2]\rangle \\ &\leqslant \text{im}\,\mu'. \end{aligned} \qquad (11.3.4)$$

The primes dividing the discriminant of C are 2, 3 and 5. By (1.16),

$$\begin{aligned} \text{im}\,\mu' &\leqslant M = \left(\mathbb{Q}(S)\right)^{\times 4} \\ &= \left(\{\pm 1, \pm 2, \pm 3, \pm 6 \pm 5, \pm 10, \pm 15, \pm 30\}\right)^{\times 4} \\ &= \langle [-1,1,1,1], [2,1,1,1], [3,1,1,1], [5,1,1,1], \\ &\quad\;\; [1,-1,1,1], [1,2,1,1], [1,3,1,1], [1,5,1,1], \\ &\quad\;\; [1,1,-1,1], [1,1,2,1], [1,1,3,1], [1,1,5,1], \\ &\quad\;\; [1,1,1,-1], [1,1,1,2], [1,1,1,3], [1,1,1,5]\rangle. \end{aligned} \qquad (11.3.5)$$

It is sufficient to show that $H = \operatorname{im} \mu'$. Since all possible 2-torsion occurs in $\mathfrak{G}$, we have without any further computation that, for any p,

$$\#\mathfrak{G}_p[2] = \#\mathfrak{G}[2] = 16. \tag{11.3.6}$$

First consider $p = \infty$, so that $\mathfrak{G}_p = \mathfrak{G}_\infty = J(\mathbb{R})$. A set of representatives for $\mathbb{R}^*/(\mathbb{R}^*)^2$ is $\{\pm 1\}$. In diagram (2.1), we can take

$$
\begin{aligned}
M_\infty &= \left(\mathbb{R}^*/(\mathbb{R}^*)^2\right)^{\times 4} = \{\pm 1\}^{\times 4} \\
&= \langle [-1,1,1,1], [1,-1,1,1], [1,1,-1,1], [1,1,1,-1]\rangle, \\
\ker j_\infty &= M \cap \left((\mathbb{R}^*)^2/(\mathbb{Q}^*)^2\right)^{\times 4} \\
&= \langle [2,1,1,1], [3,1,1,1], [5,1,1,1], [1,2,1,1], \\
&\quad\ [1,3,1,1], [1,5,1,1], [1,1,2,1], [1,1,3,1], \\
&\quad\ [1,1,5,1], [1,1,1,2], [1,1,1,3], [1,1,1,5]\rangle.
\end{aligned}
\tag{11.3.7}
$$

The points (i), ..., (v) of (3) give two independent members of M_∞, namely (i) = (ii) and (iii) = (iv) = (v). From (2.2) and (6), it follows that

$$\#\mathfrak{G}_\infty/2\mathfrak{G}_\infty = \#\mathfrak{G}_\infty[2]/4 = 16/4 = 4 \tag{11.3.8}$$

and so the known members of $\mathfrak{G}/2\mathfrak{G}$ also generate $\mathfrak{G}_\infty/2\mathfrak{G}_\infty$.

$$
\begin{aligned}
\mathfrak{G}_\infty/2\mathfrak{G}_\infty &= \langle\{(0,0),\infty\}, \{(5,0),\infty\}\rangle, \\
\mu'_\infty(\mathfrak{G}_\infty/2\mathfrak{G}_\infty) &= \langle [1,-1,-1,-1], [1,1,1,-1]\rangle \leqslant \left(\mathbb{R}^*/(\mathbb{R}^*)^2\right)^{\times 4}.
\end{aligned}
\tag{11.3.9}
$$

The commutativity of (2.1) tells us that $\operatorname{im} \mu' \leqslant j_\infty^{-1}\left(\mu'_\infty(\mathfrak{G}_\infty/2\mathfrak{G}_\infty)\right)$.

Now consider $p = 3$, so that $\mathfrak{G}_p = \mathfrak{G}_3 = J(\mathbb{Q}_3)$. A set of representatives for $\mathbb{Q}_3^*/(\mathbb{Q}_3^*)^2$ is $\{\pm 1, \pm 3\}$. Also, $\mathbb{Q}(S) \cap (\mathbb{Q}_3^*)^2 = \{1, -2, -5, 10\}$. So, in diagram (2.1), we can take

$$
\begin{aligned}
M_3 &= \left(\mathbb{Q}_3^*/(\mathbb{Q}_3^*)^2\right)^{\times 4} = \{\pm 1, \pm 3\}^{\times 4} \\
\ker j_3 &= M \cap \left((\mathbb{Q}_3^*)^2/(\mathbb{Q}^*)^2\right)^{\times 4} \\
&= \langle [-2,1,1,1], [-5,1,1,1], [1,-2,1,1], [1,-5,1,1], \\
&\quad\ [1,1,-2,1], [1,1,-5,1], [1,1,1,-2], [1,1,1,-5]\rangle.
\end{aligned}
\tag{11.3.10}
$$

In M_3, we see that (iv) = (ii)·(iii) in (3). After discarding (iv), the remaining rows (i), (ii), (iii), (v) are independent in M_3, as can be seen by writing them in terms of the representatives for M_3 given in (10). From (2.2) and (6), we know that

$$\#\mathfrak{G}_3/2\mathfrak{G}_3 = \#\mathfrak{G}_3[2] = 16 \tag{11.3.11}$$

and so again the known members of $\mathfrak{G}/2\mathfrak{G}$ of (3) are sufficient to generate all of $\mathfrak{G}_3/2\mathfrak{G}_3$. This gives

$$\mathfrak{G}_3/2\mathfrak{G}_3 = \langle \{(0,0),\infty\}, \{(1,0),\infty\}, \{(2,0),\infty\}, \{(3,6),\infty\} \rangle,$$
$$\mu_3'(\mathfrak{G}_3/2\mathfrak{G}_3) = \langle [-3,-1,1,1], [1,1,-1,-1],$$
$$[-1,1,-3,-3], [3,-1,1,1] \rangle$$
$$\leqslant \left(\mathbb{Q}_3^* / (\mathbb{Q}_3^*)^2 \right)^{\times 4}, \tag{11.3.12}$$

where of course these are just (i), (ii), (iii), (v) of (3), written in terms of the representatives for M_3 given in (10).

For $p = 5$, a set of representatives for $\mathbb{Q}_5^*/(\mathbb{Q}_5^*)^2$ is $\{1,2,5,10\}$. Also, $\mathbb{Q}(S) \cap (\mathbb{Q}_5^*)^2 = \{\pm 1, \pm 6\}$ so that

$$M_5 = \left(\mathbb{Q}_5^* / (\mathbb{Q}_5^*)^2 \right)^{\times 4} = \{1,2,5,10\}^{\times 4},$$
$$\ker j_5 = M \cap \left((\mathbb{Q}_5^*)^2 / (\mathbb{Q}^*)^2 \right)^{\times 4} \tag{11.3.13}$$
$$= \langle [-1,1,1,1], [6,1,1,1], [1,-1,1,1], [1,6,1,1],$$
$$[1,1,-1,1], [1,1,6,1], [1,1,1,-1], [1,1,1,6] \rangle.$$

In M_5, we again have that (iv) = (ii) $\cdot$ (iii) in (3). After discarding (iv), the remaining rows are independent in M_5. From (2.2) and (6), we know that

$$\#\mathfrak{G}_5/2\mathfrak{G}_5 = \#\mathfrak{G}_5[2] = 16 \tag{11.3.14}$$

and so again the known members of $\mathfrak{G}/2\mathfrak{G}$ of (3) are sufficient to generate all of $\mathfrak{G}_5/2\mathfrak{G}_5$. This gives

$$\mathfrak{G}_5/2\mathfrak{G}_5 = \langle \{(0,0),\infty\}, \{(1,0),\infty\}, \{(2,0),\infty\}, \{(3,6),\infty\} \rangle,$$
$$\mu_5'(\mathfrak{G}_5/2\mathfrak{G}_5) = \langle [10,1,2,5], [1,5,1,1], [2,1,1,2], [2,2,1,2] \rangle$$
$$\leqslant \left(\mathbb{Q}_5^* / (\mathbb{Q}_5^*)^2 \right)^{\times 4}. \tag{11.3.15}$$

Finally, for $p = 2$, a set of representatives for $\mathbb{Q}_2^*/(\mathbb{Q}_2^*)^2$ is given by $\{\pm 1, \pm 2, \pm 3, \pm 6\}$. Also, $\mathbb{Q}(S) \cap (\mathbb{Q}_2^*)^2 = \{1, -15\}$ so that

$$M_2 = \left(\mathbb{Q}_2^* / (\mathbb{Q}_2^*)^2 \right)^{\times 4} = \{\pm 1, \pm 2, \pm 3, \pm 6\}^{\times 4}$$
$$\ker j_2 = M \cap \left((\mathbb{Q}_2^*)^2 / (\mathbb{Q}^*)^2 \right)^{\times 4} \tag{11.3.16}$$
$$= \langle [-15,1,1,1], [1,-15,1,1], [1,1,-15,1], [1,1,1,-15] \rangle.$$

In M_2, we see that all of (i), $\dots$, (v) in (3) are still independent. From (2.2) and (6), we know that

$$\#\mathfrak{G}_2/2\mathfrak{G}_2 = \#\mathfrak{G}_2[2] \cdot 4 = 64 \tag{11.3.17}$$

and so we are missing precisely one further generator for $\mathfrak{G}_2/2\mathfrak{G}_2$. A short search reveals $\{(12,\gamma),\infty\}$, where $\gamma \in \mathbb{Q}_2$ and $\gamma^2 = F(12) = 12^2 \cdot 385$.¶ The existence of $\gamma \in \mathbb{Q}_2$ is guaranteed by the fact that $385 \equiv 1 \pmod 8$. Under the map μ_2' we have

$$\mu_2' : \{(12,\gamma),\infty\} \longmapsto [12,11,10,7]$$
$$= [3,3,-6,-1] \in \left(\mathbb{Q}_2^*/(\mathbb{Q}_2^*)^2\right)^{\times 4}, \tag{11.3.18}$$

and so we can see from the following set of images under μ_2' that $\mathfrak{G}_2/2\mathfrak{G}_2$ is generated by the known members of $\mathfrak{G}/2\mathfrak{G}$ given as (i),...,(v) in (3), together with $\{(12,\gamma),\infty\}$.

$$\mathfrak{G}_2/2\mathfrak{G}_2 = \langle \{(0,0),\infty\}, \{(1,0),\infty\}, \{(2,0),\infty\},$$
$$\{(5,0),\infty\}, \{(3,6),\infty\}, \{(12,\gamma),\infty\} \rangle,$$
$$\mu_2'(\mathfrak{G}_2/2\mathfrak{G}_2) = \langle [-1,-1,-2,3], [1,3,-1,-1], [2,1,6,-3], \tag{11.3.19}$$
$$[-3,1,3,1], [3,2,1,-2], [3,3,-6,-1] \rangle$$
$$\leqslant \left(\mathbb{Q}_2^*/(\mathbb{Q}_2^*)^2\right)^{\times 4},$$

For each, from the commutativity of (2.1) we know that $\operatorname{im}\mu'$ is contained in $j_p^{-1}(\operatorname{im}\mu_p')$. In particular,

$$\operatorname{im}\mu' \leqslant j_\infty^{-1}(\operatorname{im}\mu_\infty') \cap j_3^{-1}(\operatorname{im}\mu_3') \cap j_5^{-1}(\operatorname{im}\mu_5') \cap j_2^{-1}(\operatorname{im}\mu_2')$$
$$= \langle \ker j_\infty, H \rangle \cap \langle \ker j_3, H \rangle \cap \langle \ker j_5, H \rangle \tag{11.3.20}$$
$$\cap \langle \ker j_2, H, [3,3,-6,-1] \rangle.$$

We have already found generators for each of the five groups H, $\ker j_\infty$, $\ker j_3$, $\ker j_5$, $\ker j_2$, in equations (4),(7),(10),(13),(16), respectively. It is now a quick finite computation to check that the intersection in (20) is just the same as the group H of (4). This can be done in a few seconds by computer, or one can do it with pen and paper by working 'component by component'. For example, note that the fourth coordinates of the members of H span all of $\mathbb{Q}(\mathcal{S}) = \langle -1,2,3,5 \rangle$, and that the only members of H with 1 as the fourth coordinate are: $[1,1,1,1]$ and $[6,5,1,1]$. So it is sufficient to show that there are no other elements in the right hand side of (20) with 1 as the fourth coordinate. It is a quick pen and paper computation to check that the only members of both $\langle \ker j_3, H \rangle$ and $\langle \ker j_5, H \rangle$ with 1 as the

---

¶ We could equally well have chosen $x = \frac{1}{4}$ instead of $x = 12$, to obtain the missing generator. Indeed, running through $x = \frac{k}{4}$, for k odd, often seems to be a quick way of finding a missing generator of $\mathfrak{G}_2/2\mathfrak{G}_2$.

fourth coordinate, also have 1 as the third coordinate. It is then quick to check that the only members of $\langle \ker j_2, H, [3,3,-6,-1] \rangle$ with 1 as both the third and fourth components are

$$[1,1,1,1], [-15,1,1,1], [1,-15,1,1], [-15,-15,1,1],$$
$$[6,-3,1,1], [-10,-3,1,1], [6,5,1,1], [-10,5,1,1]. \tag{11.3.21}$$

But, of these, only $[1,1,1,1]$ and $[6,5,1,1]$ are in $\langle \ker j_\infty, H \rangle$. Hence the right hand side of (20) is just H. Combining this with (4) gives

$$\operatorname{im} \mu' = H, \tag{11.3.22}$$

and so $\mathfrak{G}$ has rank 1, as required.

The other example originally found by Gordon and Grant (1993) using homogeneous spaces is the following.

Example 11.3.2. *The jacobian of* $Y^2 = X(X-3)(X-4)(X-6)(X-7)$ *has rank 0, with* $\mathfrak{G}_{\text{tors}}$ *given by the 16 points of order 2.*

We leave it as an exercise for the reader to rework this example in the same style as Example 1. There is a genus 3 example in Schaefer (1995) which, in fact, was the first computation in higher genus to use the style of argument presented in this section. A genus 2 example of the form $Y^2 = F(X)$, where $F(X)$ is a sextic, irreducible over $\mathbb{Q}$, can be found in Flynn, Poonen and Schaefer (1995).

4   A worked example of descent via isogeny. This section also reworks a rank computation; in this case, a descent via isogeny on the jacobian of a curve of genus 2. This was originally due to Flynn (1994), who used computer algebra to derive explicit equations for the homogeneous spaces which arise in a complete descent via isogeny. However, we shall instead imitate the type of argument used in Example 2.1 and so, as in Section 3, avoid the use of homogeneous spaces.

Example 11.4.1. *Let* $\mathcal{C}_2$ *be the curve (8.1.12); that is*

$$\mathcal{C}_2: \quad Y^2 = G_1(X)G_2(X)G_3(X) = (X^2+1)(X^2+2)(X^2+X+1), \tag{11.4.1}$$

and let $\widehat{\mathcal{C}}_2$ *be the curve (10.2.18) given by*

$$\widehat{\mathcal{C}}_2: \quad Y^2 = L_1(X)L_2(X)L_3(X) = (X^2-2X-2)(-X^2+1)(2X), \tag{11.4.2}$$

whose jacobian is isogenous to that of C_2. Then $\mathfrak{G}_{\text{tors}} \cong C_2^{\times 2}$ and $\mathfrak{G}$ has rank 1, with $\mathfrak{G}/2\mathfrak{G}$ generated by $\mathfrak{G}_{\text{tors}}$ and $\{\infty^+, \infty^+\}$. Also, $\widehat{\mathfrak{G}}_{\text{tors}} \cong C_2^{\times 3}$ and $\widehat{\mathfrak{G}}$ has rank 1, with $\widehat{\mathfrak{G}}/2\widehat{\mathfrak{G}}$ generated by $\widehat{\mathfrak{G}}_{\text{tors}}$ and $\{(0,0), (-1/2, 3/4)\}$.

Proof. It was shown in Example 8.2.2 that $\mathfrak{G}_{\text{tors}} \cong C_2^{\times 2}$, the 2-torsion group of order 4, generated by

$$\mathfrak{A}_1 = \{(i, 0), (-i, 0)\}, \quad \mathfrak{A}_2 = \{(\sqrt{-2}, 0), (\sqrt{-2}, 0)\}. \tag{11.4.3}$$

For $\widehat{\mathfrak{G}}$, we see that reductions modulo 5 and 7 give jacobian groups of orders 24 and 64, respectively, and so $\widehat{\mathfrak{G}}_{\text{tors}} \cong C_2^{\times 3}$, the 2-torsion group of order 8, generated by

$$\widehat{\mathfrak{A}}_1 = \{(1 + \sqrt{3}, 0), (1 - \sqrt{3}, 0)\}, \quad \widehat{\mathfrak{A}}_2 = \{(1, 0), (-1, 0)\},$$
$$\widehat{\mathfrak{B}} = \{(1, 0), (0, 0)\}. \tag{11.4.4}$$

There is $\{\infty^+, \infty^+\} \in \mathfrak{G}$ of infinite order; it is not in $2\mathfrak{G}$ by Lemma 6.5.1, but is in $\hat{\phi}(\widehat{\mathfrak{G}})$. There is also

$$\widehat{\mathfrak{C}} = \{(0,0), (-\frac{1}{2}, \frac{3}{4})\} \tag{11.4.5}$$

in $\widehat{\mathfrak{G}}$ of infinite order, which is not in $\phi(\mathfrak{G})$, from (10.2.19), and so is certainly not in $2\widehat{\mathfrak{G}}$.

The set S of (1.4) is $\{2, 3\}$ and so

$$\mathbb{Q}(S) = \{\pm 1, \pm 2, \pm 3, \pm 6\}, \tag{11.4.6}$$

giving

$$\text{im}\,\mu^\phi, \ \text{im}\,\mu^{\hat{\phi}} \ \leqslant M = \{\pm 1, \pm 2, \pm 3, \pm 6\}^{\times 2}$$
$$= \langle [-1, 1], [2, 1], [3, 1], [1, -1], [1, 2], [1, 3] \rangle. \tag{11.4.7}$$

The images of the known members of $\widehat{\mathfrak{G}}/\phi(\mathfrak{G})$ under the map μ^ϕ of (10.2.8), (10.2.13) are as follows.

$$\mu^\phi : \widehat{\mathfrak{A}}_1 \longmapsto [6, -3], \ \widehat{\mathfrak{A}}_2 \longmapsto [-3, 3], \ \widehat{\mathfrak{B}} \longmapsto [6, -6], \ \widehat{\mathfrak{C}} \longmapsto [6, 3]. \tag{11.4.8}$$

So, $\widehat{H} \leqslant \text{im}\,\mu^\phi$, where

$$\widehat{H} = \langle [6, -3], [-3, 3], [6, -6], [6, 3] \rangle \leqslant (\mathbb{Q}^* / (\mathbb{Q}^*)^2)^{\times 2}. \tag{11.4.9}$$

Similarly, as computed in Example 10.2.1, the images of the known members of $\mathfrak{G}/\hat{\phi}(\widehat{\mathfrak{G}})$ under $\mu^{\hat{\phi}}$ are

$$\mu^{\hat{\phi}} : \mathfrak{A}_1 \longmapsto [1,1], \quad \mathfrak{A}_2 \longmapsto [1,3]. \tag{11.4.10}$$

So $H \leqslant \operatorname{im} \mu^{\hat{\phi}}$, where

$$H = \langle [1,3] \rangle \leqslant \left(\mathbb{Q}^*/(\mathbb{Q}^*)^2 \right)^{\times 2}. \tag{11.4.11}$$

It is sufficient to show that $\widehat{H} = \operatorname{im} \mu^{\phi}$ and $H = \operatorname{im} \mu^{\hat{\phi}}$.

For $p = \infty$, a set of representatives for $\mathbb{R}^*/(\mathbb{R}^*)^2$ is $\{\pm 1\}$. In either diagram (2.23) or the isogenous diagram, we can take

$$\begin{aligned}
M_\infty &= \left(\mathbb{R}^*/(\mathbb{R}^*)^2 \right)^{\times 2} = \{\pm 1\}^{\times 2} = \langle [-1,1], [1,-1] \rangle, \\
\ker j_\infty &= M \cap \left((\mathbb{R}^*)^2/(\mathbb{Q}^*)^2 \right)^{\times 2} \\
&= \langle [2,1], [3,1], [1,2], [1,3] \rangle.
\end{aligned} \tag{11.4.12}$$

The points $\widehat{\mathfrak{A}}_1$ and $\widehat{\mathfrak{A}}_2$ map to two independent members of M_∞ and so are themselves independent members of $\widehat{\mathfrak{G}}_\infty/\phi(\mathfrak{G}_\infty)$. From (10.4.8) we know that

$$\#\widehat{\mathfrak{G}}_\infty/\phi(\mathfrak{G}_\infty) \cdot \#\mathfrak{G}_\infty/\hat{\phi}(\widehat{\mathfrak{G}}_\infty) = (4/2)^2 = 4, \tag{11.4.13}$$

and so the known members of $\widehat{\mathfrak{G}}/\phi(\mathfrak{G})$ and $\mathfrak{G}/\hat{\phi}(\widehat{\mathfrak{G}})$ generate $\widehat{\mathfrak{G}}_\infty/\phi(\mathfrak{G}_\infty)$ and $\mathfrak{G}_\infty/\hat{\phi}(\widehat{\mathfrak{G}}_\infty)$, respectively.

$$\begin{aligned}
\widehat{\mathfrak{G}}_\infty/\phi(\mathfrak{G}_\infty) &= \langle \widehat{\mathfrak{A}}_1, \widehat{\mathfrak{A}}_2 \rangle, \\
\mu^{\phi}_\infty \left(\widehat{\mathfrak{G}}_\infty/\phi(\mathfrak{G}_\infty) \right) &= \langle [1,-1], [-1,1] \rangle \leqslant \left(\mathbb{R}^*/(\mathbb{R}^*)^2 \right)^{\times 2}, \\
\mathfrak{G}_\infty/\hat{\phi}(\widehat{\mathfrak{G}}_\infty) & \text{ is the trivial group.}
\end{aligned} \tag{11.4.14}$$

For $p = 3$, a set of representatives for $\mathbb{Q}_3^*/(\mathbb{Q}_3^*)^2$ is $\{\pm 1, \pm 3\}$. Furthermore, $M \cap (\mathbb{Q}_3^*)^2 = \{1, -2\}$. In either diagram (2.23) or the isogenous diagram, we can take

$$\begin{aligned}
M_3 &= \left(\mathbb{Q}_3^*/(\mathbb{Q}_3^*)^2 \right)^{\times 2} = \left(\{\pm 1, \pm 3\} \right)^{\times 2} = \langle [-1,1], [3,1], [1,-1], [1,3] \rangle, \\
\ker j_3 &= M \cap \left((\mathbb{Q}_3^*)^2/(\mathbb{Q}^*)^2 \right)^{\times 2} \\
&= \langle [-2,1], [1,-2] \rangle.
\end{aligned} \tag{11.4.15}$$

The points $\widehat{\mathfrak{A}}_1$ and $\widehat{\mathfrak{A}}_2$ map to 2 independent members of M_3 and so are themselves independent members of $\widehat{\mathfrak{G}}_3/\phi(\mathfrak{G}_3)$. Note however that we have $[-3,3] = [6,-6] = [6,3]$ in M_3 and so $\widehat{\mathfrak{A}}_2 = \widehat{\mathfrak{B}} = \widehat{\mathfrak{C}} \in \widehat{\mathfrak{G}}_3/\phi(\mathfrak{G}_3)$.

Also, $[1,3] \neq [1,1]$ in M_3 so that $\mathfrak{A}_2$ is a nontrivial member of $\mathfrak{G}_3/\hat{\phi}(\widehat{\mathfrak{G}}_3)$. From (10.4.8) we know that

$$\#\widehat{\mathfrak{G}}_3/\phi(\mathfrak{G}_3) \cdot \#\mathfrak{G}_3/\hat{\phi}(\widehat{\mathfrak{G}}_3) = 4^2 = 16, \qquad (11.4.16)$$

The known members of $\widehat{\mathfrak{G}}/\phi(\mathfrak{G})$ and $\mathfrak{G}/\hat{\phi}(\widehat{\mathfrak{G}})$ generate groups of order 4 and 2 in $\widehat{\mathfrak{G}}_3/\phi(\mathfrak{G}_3)$ and $\mathfrak{G}_3/\hat{\phi}(\widehat{\mathfrak{G}}_3)$, respectively. We are still missing exactly one generator for either $\widehat{\mathfrak{G}}_3/\phi(\mathfrak{G}_3)$ or $\mathfrak{G}_3/\hat{\phi}(\widehat{\mathfrak{G}}_3)$. A short search reveals

$$\mathfrak{D} = \{(\beta, 0), \infty^+\} \in \mathfrak{G}_3, \quad \beta \in \mathbb{Q}_3, \quad \beta^2 = -2, \quad \beta \equiv 1 \pmod 3. \quad (11.4.17)$$

This maps to $[2, -3]$ under $\mu^{\hat{\phi}}$ and so is independent of $\mathfrak{A}_2$ in $\mathfrak{G}/\hat{\phi}(\widehat{\mathfrak{G}})$. We now have, using the generators for M_3 of (15),

$$\begin{aligned}
\widehat{\mathfrak{G}}_3/\phi(\mathfrak{G}_3) &= \langle \widehat{\mathfrak{A}}_1, \widehat{\mathfrak{A}}_2 \rangle, \\
\mu_3^\phi(\widehat{\mathfrak{G}}_3/\phi(\mathfrak{G}_3)) &= \langle [-3, -3], [-3, 3] \rangle \leqslant (\mathbb{Q}_3^*/(\mathbb{Q}_3^*)^2)^{\times 2}, \\
\mathfrak{G}_3/\hat{\phi}(\widehat{\mathfrak{G}}_3) &= \langle \mathfrak{A}_2, \mathfrak{D} \rangle, \\
\mu_3^{\hat{\phi}}(\mathfrak{G}_3/\hat{\phi}(\widehat{\mathfrak{G}}_3)) &= \langle [1, 3], [2, -3] \rangle \leqslant (\mathbb{Q}_3^*/(\mathbb{Q}_3^*)^2)^{\times 2}.
\end{aligned} \qquad (11.4.18)$$

Finally, for $p = 2$, a set of representatives for $\mathbb{Q}_2^*/(\mathbb{Q}_2^*)^2$ is given by $\mathbb{Q}(S) = \{\pm 1, \pm 2, \pm 3, \pm 6\}$, and $M \cap (\mathbb{Q}_3^*)^2$ is the trivial group. In either diagram (2.23) or the isogenous diagram, we can take

$$\begin{aligned}
M_2 &= (\mathbb{Q}_2^*/(\mathbb{Q}_2^*)^2)^{\times 2} = (\{\pm 1, \pm 2, \pm 3 \pm 6\})^{\times 2} \\
&= \langle [-1, 1], [2, 1], [3, 1], [1, -1], [1, 2], [1, 3] \rangle, \qquad (11.4.19) \\
\ker j_2 &= M \cap ((\mathbb{Q}_2^*)^2/(\mathbb{Q}^*)^2)^{\times 2} = \text{ the trivial group.}
\end{aligned}$$

Since j_2 is injective, the points $\widehat{\mathfrak{A}}_1$, $\widehat{\mathfrak{A}}_2$, $\widehat{\mathfrak{B}}$ and $\widehat{\mathfrak{C}}$ map to four independent members of M_2 and so are themselves independent members of $\widehat{\mathfrak{G}}_2/\phi(\mathfrak{G}_2)$. Similarly, $\mathfrak{A}_2$ must give a nontrivial member of $\mathfrak{G}_2/\hat{\phi}(\widehat{\mathfrak{G}}_2)$. From (10.4.8) we know that

$$\#\widehat{\mathfrak{G}}_2/\phi(\mathfrak{G}_2) \cdot \#\mathfrak{G}_2/\hat{\phi}(\widehat{\mathfrak{G}}_2) = (4 \cdot 2)^2 = 64, \qquad (11.4.20)$$

The known members of $\widehat{\mathfrak{G}}/\phi(\mathfrak{G})$ and $\mathfrak{G}/\hat{\phi}(\widehat{\mathfrak{G}})$ generate groups of order 16 and 2 in $\widehat{\mathfrak{G}}_2/\phi(\mathfrak{G}_2)$ and $\mathfrak{G}_2/\hat{\phi}(\widehat{\mathfrak{G}}_2)$, respectively, and so we are again missing exactly one generator. A short search reveals

$$\widehat{\mathfrak{C}} = \{(-11, 12\gamma), \infty\} \in \widehat{\mathfrak{G}}_2, \quad \gamma \in \mathbb{Q}_2, \quad \gamma^2 = 2585. \qquad (11.4.21)$$

This maps to $[6, 2]$ under μ^ϕ and so gives the remaining generator.

$$\widehat{\mathfrak{G}}_2/\phi(\mathfrak{G}_2) = \langle \widehat{\mathfrak{A}}_1, \widehat{\mathfrak{A}}_2, \widehat{\mathfrak{B}}, \widehat{\mathfrak{C}}, \widehat{\mathfrak{E}} \rangle,$$

$$\mu_2^\phi\left(\widehat{\mathfrak{G}}_2/\phi(\mathfrak{G}_2)\right) = \langle [6, -3], [-3, 3], [6, -6], [6, 3], [6, 2] \rangle \leqslant \left(\mathbb{Q}_2^*/(\mathbb{Q}_2^*)^2\right)^{\times 2},$$

$$\mathfrak{G}_2/\hat{\phi}(\widehat{\mathfrak{G}}_2) = \langle \mathfrak{A}_2 \rangle,$$

$$\mu_2^{\hat{\phi}}\left(\mathfrak{G}_2/\hat{\phi}(\widehat{\mathfrak{G}}_2)\right) = \langle [1, 3] \rangle \leqslant \left(\mathbb{Q}_2^*/(\mathbb{Q}_2^*)^2\right)^{\times 2}. \tag{11.4.22}$$

The determination of $\widehat{\mathfrak{G}}/\phi(\mathfrak{G})$ and $\mathfrak{G}/\hat{\phi}(\widehat{\mathfrak{G}})$ is now straightforward. We first observe that diagram (2.23) with $p = 3$ gives that

$$\operatorname{im}\mu^\phi \leqslant j_3^{-1}\left(\operatorname{im}\mu_3^\phi\right) = \langle \ker j_3, \widehat{H} \rangle = \widehat{H}, \tag{11.4.23}$$

and so we can deduce from (9) that

$$\operatorname{im}\mu^\phi = \widehat{H} \tag{11.4.24}$$

without requiring any of the information at $\mathbb{Q}_2$ or $\mathbb{R}$.

Similarly, we can consider the diagram isogenous to (2.23), that is, with $\widehat{\mathfrak{G}}/\phi(\mathfrak{G})$, μ^ϕ, $\widehat{\mathfrak{G}}_p/\phi(\mathfrak{G}_p)$ and μ_p^ϕ replaced by $\mathfrak{G}/\hat{\phi}(\widehat{\mathfrak{G}})$, $\mu^{\hat{\phi}}$, $\mathfrak{G}_p/\hat{\phi}_p(\widehat{\mathfrak{G}}_p)$ and $\mu_p^{\hat{\phi}}$, respectively. When $p = 2$, we have that

$$\operatorname{im}\mu^{\hat{\phi}} \leqslant j_2^{-1}\left(\operatorname{im}\mu_2^{\hat{\phi}}\right) = \langle \ker j_2, H \rangle = H, \tag{11.4.25}$$

and so we can deduce from (11) that $\operatorname{im}\mu^{\hat{\phi}} = H$, without requiring any of the information at $\mathbb{Q}_3$ or $\mathbb{R}$. We have now shown that both $\widehat{\mathfrak{G}}/\phi(\mathfrak{G})$ and $\mathfrak{G}/\hat{\phi}(\widehat{\mathfrak{G}})$ are generated by the elements given in (8) and (10), respectively, as required.

In an entirely different spirit, there are also the rank computations in Mazur (1977) relating to Eisenstein quotients, and Brumer (1995a).

We should mention that the techniques we have illustrated here are essentially methods of confirmation. In each example, there were easy-to-find small $\mathbb{Q}$-rational points which turned out to generate $\mathfrak{G}/2\mathfrak{G}$, and the technique gave a quick way of proving, in each case, that the rank was indeed what we hoped it would be. We have sidestepped the question of how in general to search for missing generators of $\mathfrak{G}/2\mathfrak{G}$. There is of course no known algorithm for doing this, but for elliptic curves there are techniques which at least make the process of searching up to some bound more efficient, by using homogeneous spaces [see Cremona (1992)]. The situation in genus 2 is still somewhat unsatisfactory. There are models for homogeneous spaces [see Gordon & Grant (1993)], but the size (and number) of the equations is large and at the moment searching for $\mathbb{Q}$-rational points on them seems to be a slow process.

5   Large rank. It is only very recently that any serious attempt has been made to search for curves of genus 2 whose jacobians have large rank, and there is nothing like the substantial elliptic curves literature [see, for example, Mestre (1991c) and Nagao & Kouya (1994)]. The best genus 2 result so far is in Stoll (1995), where two curves, defined over $\mathbb{Q}$, are found whose jacobians each have rank at least 19. One of these curves is

$$
\begin{aligned}
Y^2 &= F(X) \\
&= 1\,960\,641 X^6 - 14\,210\,996 X^5 - 149\,332\,238 X^4 + 1\,238\,887\,722 X^3 \\
&\quad + 2\,145\,729\,513 X^2 - 23\,268\,170\,226 X + 49\,641\,176\,809. \quad (11.5.1)
\end{aligned}
$$

The other curve has integer coefficients of similar size. To show results of this type (where the rank is shown to be at least some value but is not computed exactly), it is not necessary to imitate the computations of Sections 3, 4. It is sufficient to find 19 non-torsion members of $\mathfrak{G} = J(\mathbb{Q})$ and show that they are independent of each other and any torsion, using the map μ from $\mathfrak{G}$ to $\mathbb{Q}[\Theta]^*/\mathbb{Q}^*\{\mathbb{Q}[\Theta]^*\}^2$ described in (6.1.9) [recall that Θ represents a root of $F(X)$ which, in this case, is irreducible over $\mathbb{Q}$)] and Lemma 6.5.1. The curve (1) has at least the 20 rational points $P_j = (x_j, y_j)$, where

$$
\begin{aligned}
(x_0, \ldots, x_{19}) = \Big(&-16, -12, -10, -8, -5, -2, 0, 1, 3, 4, 6, 10, 11, 14, \\
&\frac{151}{2}, -\frac{397}{4}, \frac{59}{23}, -\frac{89}{27}, \frac{269}{32}, -\frac{9153}{373} \Big) \quad (11.5.2)
\end{aligned}
$$

and each y_j is taken to be the positive square root of $F(x_j)$.

First note that $\mathfrak{G}$ has no torsion, as can be seen from $\tilde{J}(\#\mathbb{F}_{11}) = 252$ and $\tilde{J}(\#\mathbb{F}_{17}) = 625$, which are coprime. One can check that the 19 divisors $\{P_0, P_j\}$¶ map to 18 independent members of $\mathbb{Q}[\Theta]^*/\mathbb{Q}^*\{\mathbb{Q}[\Theta]^*\}^2$ under μ, and so the rank of $\mathfrak{G}$ is at least 18. This fact can be established either by direct computation in $\mathbb{Q}[\Theta]^*/\mathbb{Q}^*\{\mathbb{Q}[\Theta]^*\}^2$ or, more simply, by using information from reductions modulo choices of prime p such that $p \nmid 2\,\mathrm{disc}\,(F)$ and $\tilde{F}(X)$ splits completely over $\mathbb{F}_p$ (an example is $p = 2027$).

¶ In fact, Stoll looks at the divisors $P_j - P_0$, which are equivalent to the divisors $\{P_0, P_j\} - \{P_0, P_0\}$ in our notation. But the images under μ are the same, since of course $\{P_0, P_0\}$ is taken to be the identity.

This is described in [Stoll (1995), pp. 1342–1343]. Finally, Lemma 6.5.1 can be used to show that $\mathfrak{W}$ is not contained in $\mathfrak{G}$ [Criterion (i) clearly fails since $F(X)$ has no $\mathbb{Q}$-rational roots, and one can construct the polynomial $h(X)$ of (6.5.4), using the file **kernel.of.mu**, available by anonymous ftp, if desired; $h(X)$ is easily shown to have no $\mathbb{Q}$-rational roots, and so Criterion (ii) fails also]. It follows that the divisors of the form $\{P_j, P_j\}$ lie in the kernel of μ but not in $2\mathfrak{G}$. Therefore $\{P_0, P_0\}$,¶ say, gives an additional member of $\mathfrak{G}$ independent from the other 18.†

It is apparent that the skill is not so much in the verification, but in the finding of such examples as (1). The curve would certainly certainly not have been found using a crude search through all curves with integer coefficients up to some bound; it is necessary to search selectively among curves whose jacobians are more likely to have large rank. The technique used to find (1) [described on page 1344 of Stoll (1995), imitating the elliptic curves technique in Mestre (1991c)] is to run through 14-tuples of rational numbers $\alpha_1, \dots, \alpha_{14}$. One constructs

$$P(X) = \prod_{k=1}^{14} (X - \alpha_k) \tag{11.5.3}$$

and then finds $H(X)$, monic of degree 7, and $F(X)$ of degree at most 6, such that

$$P(X) = H(X)^2 - F(X). \tag{11.5.4}$$

The curve $Y^2 = F(X)$ then has at least the 14 pairs $(\alpha_i, \pm H(\alpha_i))$ which generically generate a subgroup of rank at least 13 in the jacobian.‡ One can therefore search among jacobians, each of which is virtually guaranteed to have rank at least 13 as a foundation; the hope is that some of these will contain further independent generators. Note that this technique emphasizes members of $\mathfrak{G}$ of the form $\{P, Q\}$, where P, Q are both in $\mathcal{C}(\mathbb{Q})$. It would be interesting to know whether any further mileage could be gained from an

---

¶ Recall that, as usual, this stands for $P_0 + P_0 - \infty^+ - \infty^-$. In this situation, it is equivalent modulo $2\mathfrak{G}$, and somewhat simpler, to refer to the divisor $(x_0, y_0) - (x_0, -y_0)$.

† In fact, Stoll has mentioned to us that there is a slight error in the exposition of Stoll (1995). The result that the rank $\geqslant 19$ is correct, but in the paper it was claimed that this arose from 19 divisors with independent images under μ. In fact (as we have outlined here) there are only 18 independent images under μ, and the 19th independent divisor lies in the kernel of μ.

‡ Generically, the image of this subgroup under μ has rank 12, and $\mathfrak{W}$ is not contained in $2\mathfrak{G}$, giving a 13th independent member of $\mathfrak{G}$.

adapted version of the technique, emphasising members of the form $\{P, P'\}$, where P is defined over a quadratic number field.

Care should be taken to distinguish genuine genus 2 examples from reducible examples, which cheat by using pairs of elliptic curves. After all, we could consider an elliptic curve $\mathcal{E}$ of largest known rank, namely (we think) that of rank at least 21 in Nagao & Kouya (1994). Then we could find a curve of genus 2 whose jacobian is isogenous to $\mathcal{E} \times \mathcal{E}$ (how to do this in general for any $\mathcal{E}_1 \times \mathcal{E}_2$ is described in Chapter 14, Section 3), which will have rank at least 42. In fact [Stoll (1995), pp. 1343–1344] the curve (1) is a legitimate example, since the jacobian is not isogenous over $\overline{\mathbb{Q}}$ to the product of two elliptic curves. We shall say more about reducibility in Chapter 14.

Chapter 12

Heights

0 Introduction. When the ground field is a number field, we can define a height function on the jacobian. Standard properties of the height function allow us to deduce that $\mathfrak{G}$ is finitely generated from the fact (Chapter 11) that $\mathfrak{G}/2\mathfrak{G}$ is finite. Indeed, effectively computable constants relating to the height function give a way of finding generators for $\mathfrak{G}$ once $\mathfrak{G}/2\mathfrak{G}$ has been found. In practice, this takes a long time on the jacobian of a curve of genus 2. We give a recent improvement using Richelot's isogeny which allows generators to be computed for a few examples with small coefficients.

1 A height function on $\mathfrak{G}$. For any point $\mathbf{x} = (x_i) \in \mathbb{P}^n(\mathbb{Q})$, we can choose a representative (x_i) for which

$$x_i \in \mathbb{Z}, \quad \gcd(x_0, \dots, x_n) = 1, \tag{12.1.1}$$

and then define

$$H(\mathbf{x}) = \max_i |x_i|. \tag{12.1.2}$$

This is the standard height function on $\mathbb{P}^n(\mathbb{Q})$. Similarly, if we let $\mathbb{M}^n(\mathbb{Q})$ represent all $(n+1) \times (n+1)$ matrices modulo scalar multiplication, and take $M = (M_{ij}) \in \mathbb{M}^n(\mathbb{Q})$, then we can choose a representative (M_{ij}) for which

$$M_{ij} \in \mathbb{Z}, \quad \gcd(M_{ij}) = 1, \tag{12.1.3}$$

and then define

$$H(M) = (n+1) \max_{ij} |M_{ij}|. \tag{12.1.4}$$

For any $\mathbf{x}, \mathbf{y} \in \mathbb{P}^n(\mathbb{Q})$, we can construct the matrix $(x_i y_j + y_i x_j) \in \mathbb{M}^n(\mathbb{Q})$. For future reference, we recall the following technical lemma. The case $n = 2$ follows from the slightly stronger Lemma 2 on page 79 of Cassels (1991), and the generalization to any n is straightforward.

$$H(\mathbf{x})H(\mathbf{y}) \leqslant 2H\big((x_i y_j + y_i x_j)\big). \tag{12.1.5}$$

As usual, let $\mathcal{E}$ be an elliptic curve

$$\mathcal{E}: \qquad Y^2 = f_0 + f_1 X + f_2 X^2 + X^3, \tag{12.1.6}$$

and $\mathcal{C}$ be a curve of genus 2

$$\mathcal{C}: \quad Y^2 = f_0 + f_1 X + f_2 X^2 + f_3 X^3 + f_4 X^4 + f_5 X^5 + f_6 X^6, \tag{12.1.7}$$

where for simplicity we shall assume in either case that the ground field is $\mathbb{Q}$. The generalization to number fields will be outlined at the end of Section 2. As usual, we can if necessary adjust X, Y by constants so that all f_i are in $\mathbb{Z}$. Let J denote the elliptic curve $\mathcal{E}$, in the first case, or the jacobian of $\mathcal{C}$ in the second case. As usual, let $\mathfrak{G}$ denote $J(\mathbb{Q})$. For any $\mathfrak{A} \in \mathfrak{G}$, let $\big(z_i(\mathfrak{A})\big)$ represent the member of $\mathbb{P}^{4^g-1}(\mathbb{Q})$ given by the $z_0, \ldots, z_3$ of $(7.1.2)$ in the elliptic curve case, and by the coordinates $z_0, \ldots, z_{15}$ described at the beginning of Chapter 2, Section 2 in the genus 2 case. In either case, we can define a function

$$H_J: \quad \mathfrak{G} \longrightarrow \mathbb{Z}^+, \quad \mathfrak{A} \longmapsto H\Big(\big(z_i(\mathfrak{A})\big)\Big), \tag{12.1.8}$$

where H is as in (2). A simpler function can be obtained by using the Kummer coordinates $\xi = \big(\xi_i(\mathfrak{A})\big) \in \mathbb{P}^{2^g-1}(\mathbb{Q})$. In the elliptic curve case, take $\big(\xi_i(\mathfrak{A})\big) = \big(\begin{smallmatrix}1\\x\end{smallmatrix}\big) \in \mathbb{P}^1(\mathbb{Q})$, the projective x-coordinate. On the jacobian of a curve of genus 2, we can take $\big(\xi_i(\mathfrak{A})\big) \in \mathbb{P}^3(\mathbb{Q})$ to be the four projective coordinates for the Kummer surface defined in $(3.1.3)$. We can now define

$$H_\kappa: \quad \mathfrak{G} \longrightarrow \mathbb{Z}^+, \quad \mathfrak{A} \longmapsto H\Big(\big(\xi_i(\mathfrak{A})\big)\Big). \tag{12.1.9}$$

It is straightforward to deduce from the defining equations of the jacobian (given by quadratic forms) that $H_J(\mathfrak{A})$ is the same as $H_\kappa(\mathfrak{A})^2$ up to a multiplicative constant. Since we shall only be interested in these objects up to multiplicative constant, we can say that H_κ gives as much information as H_J, and so we may as well deal exclusively with H_κ since it is computationally simpler.

We first note that, for any constant C, there are only finitely many arrays of 2^g integers all with absolute value less than C, and so finitely many $\mathbf{x} \in \mathbb{P}^{2^g-1}(\mathbb{Q})$ with $H(\mathbf{x}) \leqslant C$. For each $\mathfrak{A} \in \mathfrak{G}$ there is a point $\big(\xi_i(\mathfrak{A})\big) \in \mathbb{P}^{2^g-1}(\mathbb{Q})$, as above; in the reverse direction, any $\mathbf{x} \in \mathbb{P}^{2^g-1}(\mathbb{Q})$ has at most two $\mathfrak{A} \in \mathfrak{G}$ such that $\mathbf{x} = \big(\xi_i(\mathfrak{A})\big)$. It follows that

$$\{\mathfrak{A} \in \mathfrak{G} : H_\kappa(\mathfrak{A}) \leqslant C\} \text{ is finite.} \tag{12.1.10}$$

Recall from Theorem 3.4.1 that there are polynomials B_{ij} biquadratic in the $\xi_j(\mathfrak{A})$, $\xi_j(\mathfrak{B})$ and defined over $\mathbb{Z}$ such that

$$\big(\xi_i(\mathfrak{A}+\mathfrak{B})\xi_j(\mathfrak{A}-\mathfrak{B}) + \xi_i(\mathfrak{A}-\mathfrak{B})\xi_j(\mathfrak{A}+\mathfrak{B})\big) = 2\big(B_{ij}(\mathfrak{A},\mathfrak{B})\big). \quad (12.1.11)$$

For elliptic curves these are given in Chapter 17, Lemma 4 and Chapter 24, Lemma 3 of Cassels (1991). In the genus 2 case, the construction was described in Chapter 3, and we have made the equations available by anonymous ftp as described in Appendix II. First note that, since each B_{ij} is a biquadratic form defined over $\mathbb{Z}$, we can define

$$H(B_{ij}) = \text{ sum of the absolute values of the coefficients of } B_{ij}, \quad (12.1.12)$$

in which case

$$H\Big(\big(B_{ij}(\mathfrak{A},\mathfrak{B})\big)\Big) \leqslant 2^g \cdot \max_{ij} H(B_{ij}) \cdot H(\mathfrak{A})^2 H(\mathfrak{B})^2. \quad (12.1.13)$$

Combining this with equations (5) and (11) gives that

$$H_\kappa(\mathfrak{A}+\mathfrak{B})H_\kappa(\mathfrak{A}-\mathfrak{B}) \leqslant C_1 H_\kappa(\mathfrak{A})^2 H_\kappa(\mathfrak{B})^2, \quad (12.1.14)$$

where

$$C_1 = 4 \cdot 2^g \cdot \max_{ij} H(B_{ij}). \quad (12.1.15)$$

Note that C_1 is dependent only on the coefficients f_i of the original curve, is effectively computable, and is independent of $\mathfrak{A}$ and $\mathfrak{B}$. Since any height is in $\mathbb{Z}^+$, immediate corollaries of (14) are

$$H_\kappa(\mathfrak{A}+\mathfrak{B}), H_\kappa(\mathfrak{A}-\mathfrak{B}) \leqslant C_1 H_\kappa(\mathfrak{A})^2 H_\kappa(\mathfrak{B})^2, \quad (12.1.16)$$

and

$$H_\kappa(2\mathfrak{A}) \leqslant C_1 H_\kappa(\mathfrak{A})^4. \quad (12.1.17)$$

Recall from (3.5.1) that in genus 2

$$\xi_j(2\mathfrak{A}) = B_{j4}(\mathfrak{A},\mathfrak{A}) \ (1 \leqslant j \leqslant 3), \qquad \xi_4(2\mathfrak{A}) = B_{44}(\mathfrak{A},\mathfrak{A})/2. \quad (12.1.18)$$

The same thing is true for an elliptic curve, but with $B_{14}, B_{24}, B_{34}, B_{44}$ replaced by B_{12}, B_{22}. Let $\mathbf{v} = (v_i)$, where the v_i are 2^g indeterminates, and denote

$$N(\mathbf{v}) = \big(N_j(\mathbf{v})\big) \quad (12.1.19)$$

where

$$N_j(\mathbf{v}) = B_{j4}(\mathbf{v},\mathbf{v}) \ (1 \leqslant j \leqslant 3), \qquad N_4(\mathbf{v}) = B_{44}(\mathbf{v},\mathbf{v})/2. \quad (12.1.20)$$

Similarly define N_1, N_2 to get the duplication law on the projective x-coordinate in the elliptic curve case. In either case, we have

$$N((\xi_j(\mathfrak{A}))) = (\xi_j(2\mathfrak{A})). \qquad (12.1.21)$$

Each N_j is a quartic form in the v_i, which gives an alternative way of showing (17). It is more difficult to find a constant for which the inequality of (17) can be reversed. One way to proceed is to use Hilbert's Nullstellensatz on the quartics N_j. Since they have no common zero, this tells us that, for each i, some power of v_i lies in the ideal generated by the N_j. It follows that, for some integers m, r, there exist forms $R_{ij} = R_{ij}(\mathbf{v})$ of degree r in the v_i over $\mathbb{Z}$ such that, for each i,

$$m \cdot v_i^{r+4} = \sum_j R_{ij} N_j. \qquad (12.1.22)$$

The R_{ij} may in principle be constructed explicitly by resultant computations [described in Masser and Wüstholz (1983)]. Let $H(R_{ij})$ be defined in the same way as $H(B_{ij})$ in (12) and let

$$C_2 = 2^9 \max_{ij} H(R_{ij}). \qquad (12.1.23)$$

Evaluating at

$$v_j = \xi_j(\mathfrak{A}), \quad v_j \in \mathbb{Z}, \quad \gcd_j v_j = 1, \qquad (12.1.24)$$

gives, by definition, that

$$H_\kappa(\mathfrak{A}) = \max_j |v_j|, \quad H_\kappa(2\mathfrak{A}) = \max_j |N_j| / \gcd_j N_j. \qquad (12.1.25)$$

The fact that each $R_{ij} = R_{ij}(\mathbf{v})$ is a form of degree r in the v_i, and the definition of C_2, give, for any i, that

$$\left| \sum_j R_{ij} N_j \right| \leqslant C_2 \max_j |v_j|^r \max_j |N_j| \qquad (12.1.26)$$

Combining this with (22) and cancelling $\max_j |v_j|^r$ give

$$m \cdot \max_j |v_j|^4 \leqslant C_2 \max_j |N_j|. \qquad (12.1.27)$$

But from (22) and (24), we have that the gcd of the N_j is at most m. Substituting (25) into (27) therefore gives

$$m \cdot H_\kappa(\mathfrak{A})^4 \leqslant C_2 \max_j |N_j| \leqslant C_2 \cdot m \cdot H_\kappa(2\mathfrak{A}) \qquad (12.1.28)$$

and so

$$H_\kappa(2\mathfrak{A}) \geqslant H_\kappa(\mathfrak{A})^4/C_2, \qquad (12.1.29)$$

which was our desired result. Equations (17) and (29) tell us that $H_\kappa(2\mathfrak{A})$ is the same as $H(\mathfrak{A})^4$ up to multiplicative constants which depend on the original curve, but not on $\mathfrak{A}$. It is also possible to apply a similar style of argument to obtain a constant for which inequality (14) can be reversed.

It is sometimes convenient (although we do not require it in this chapter) to define the logarithmic height

$$h(\mathfrak{A}) = \log H_\kappa(\mathfrak{A}), \qquad (12.1.30)$$

in which case the above inequalities can be written additively.

2   The Mordell-Weil Theorem.

We first recall the three key results from the previous section, namely (1.10),(1.14) and (1.29), as follows.

(i)   For any C, $\{\mathfrak{A} \in \mathfrak{G} : H_\kappa(\mathfrak{A}) \leqslant C\}$ is finite.
(ii) There exists C_1 such that for any $\mathfrak{A}, \mathfrak{B} \in \mathfrak{G}$,
$$H_\kappa(\mathfrak{A} + \mathfrak{B})H_\kappa(\mathfrak{A} - \mathfrak{B}) \leqslant C_1 H_\kappa(\mathfrak{A})^2 H_\kappa(\mathfrak{B})^2. \qquad (12.2.1)$$
(iii) There exists C_2 such that for any $\mathfrak{A} \in \mathfrak{G}$,
$$H_\kappa(2\mathfrak{A}) \geqslant H_\kappa(\mathfrak{A})^4/C_2.$$

We know from Chapter 11 that $\mathfrak{G}/2\mathfrak{G}$ is finite; say that

$$\mathfrak{G}/2\mathfrak{G} = \mathcal{T} = \{\mathfrak{B}_1, \ldots \mathfrak{B}_s\}. \qquad (12.2.2)$$

We shall imitate the argument given on p. 200 of Silverman (1986). Let $\mathfrak{A} = \mathfrak{A}_1 \in \mathfrak{G}$. Then

$$\mathfrak{A}_1 = \mathfrak{B}_{i_1} + 2\mathfrak{A}_2 \qquad (12.2.3)$$

for some $\mathfrak{B}_{i_1} \in \mathcal{T}$ and $\mathfrak{A}_2 \in \mathfrak{G}$. From (iii) and (ii) of (1)

$$\begin{aligned}
H(\mathfrak{A}_2)^4 \leqslant C_2 H(2\mathfrak{A}_2) = C_2 H(\mathfrak{A}_1 - \mathfrak{B}_{i_1}) &\leqslant C_1 C_2 H(\mathfrak{A}_1)^2 H(\mathfrak{B}_{i_1})^2 \\
&= H(\mathfrak{A}_1)^4 \big(\sqrt{C_1 C_2} H(\mathfrak{B}_{i_1})/H(\mathfrak{A}_1)\big)^2,
\end{aligned} \qquad (12.2.4)$$

and so

$$H(\mathfrak{A}_2) < H(\mathfrak{A}_1) \qquad (12.2.5)$$

whenever

$$H(\mathfrak{A}_1) > C_3, \quad \text{where } C_3 = \sqrt{C_1 C_2} \max_{1 \leqslant i \leqslant s} H(\mathfrak{B}_i). \qquad (12.2.6)$$

If also $H(\mathfrak{A}_2) > C_3$, then we can repeat this process to obtain

$$\mathfrak{A}_2 = \mathfrak{B}_{i_2} + 2\mathfrak{A}_3, \quad H(\mathfrak{A}_2) > H(\mathfrak{A}_3). \tag{12.2.7}$$

If, on continuing to repeat this process, it happened that $H(\mathfrak{A}_n) > C_3$ for all n, then there would be the contradiction that $H(\mathfrak{A}_1), H(\mathfrak{A}_2), \ldots$ is an infinite sequence of decreasing positive integers. Therefore, for some n (including the possibility $n = 1$ in the case when $H(\mathfrak{A}) \leqslant C_3$), we must have

$$H(\mathfrak{A}_n) \leqslant \sqrt{C_1 C_2} \max_{1 \leqslant i \leqslant s} H(\mathfrak{B}_i). \tag{12.2.8}$$

On recursively substituting

$$\mathfrak{A}_1 = \mathfrak{B}_{i_1} + 2\mathfrak{A}_2 = \mathfrak{B}_{i_1} + 2(\mathfrak{B}_{i_2} + 2\mathfrak{A}_3) = \ldots \tag{12.2.9}$$

we can express our original $\mathfrak{A}$ in terms of $\mathfrak{B}_{i_1}, \ldots, \mathfrak{B}_{i_{n-1}}$ and $\mathfrak{A}_n$. It follows that any $\mathfrak{A} \in \mathfrak{G}$ can be expressed as a finite sum of members of T and the set

$$\{\mathfrak{C} : H(\mathfrak{C}) \leqslant C_3\}. \tag{12.2.10}$$

But by part (i) of (1), this set is finite, and so $\mathfrak{G}$ is finitely generated.

When the ground field is a number field k, we can extend the height of (1.2) on $\mathbb{P}^n(\mathbb{Q})$ to $\mathbb{P}^n(k)$ by defining

$$H(\mathbf{x}) = \Big(\prod_v \max_{0 \leqslant i \leqslant n} |x_i|_v\Big)^{1/[k:\mathbb{Q}]}, \tag{12.2.11}$$

where $\prod_v$ is over all valuations on k which extend the usual valuations $| \; |_p$ and $| \; |_\infty$ on $\mathbb{Q}$. If $M = (M_{ij}) \in \mathbb{M}^n(k)$ then we can similarly define

$$H(M) = \Big(\prod_v \max_{i,j} |M_{ij}|_v\Big)^{1/[k:\mathbb{Q}]}. \tag{12.2.12}$$

Let $\mathfrak{G} = J(k)$. For any $\mathfrak{A} \in \mathfrak{G}$ we can now define $H_\kappa(\mathfrak{A})$ exactly as in (1.9). It can be shown, for example see p. 213 of Silverman (1986), that for any constant C, there are only finitely many $\mathbf{x} \in \mathbb{P}^n(k)$ such that $H(\mathbf{x}) \leqslant C$. As before, this gives property (i) of (1) above.

An imitation of the argument in Section 1 gives that property (ii) is satisfied with, for example,

$$C_1 = 4 \cdot 2^g (n+1) \cdot \Big(\prod_v \max_{i,j} H_v(B_{ij})\Big)^{1/[k:\mathbb{Q}]}, \tag{12.2.13}$$

where $H_v(B_{ij})$ can be chosen as

$$(\#\text{terms in } B_{ij}) \cdot \max\{|c|_v : c \text{ is a coefficient of } B_{ij}\}.$$

Note that H_κ is no longer a map into $\mathbb{Z}^+$ and so (1.16) is no longer true. However, we can replace this with

$$H_\kappa(\mathfrak{A} + \mathfrak{B}),\ H_\kappa(\mathfrak{A} - \mathfrak{B}) \leqslant (C_1/\epsilon) H_\kappa(\mathfrak{A})^2 H_\kappa(\mathfrak{B})^2, \qquad (12.2.14)$$

where

$$\epsilon = \min\{H_\kappa(\mathfrak{A}) : \mathfrak{A} \in \mathfrak{G}\}. \qquad (12.2.15)$$

Similarly, property (iii) is satisfied with

$$C_2 = 2^g(n+1) \cdot \left(\prod_v \max_{i,j} H_v(R_{ij})\right), \qquad (12.2.16)$$

where $H_v(R_{ij})$ is defined in the same way as $H_v(B_{ij})$ in (13). Imitating the argument at the beginning of this section, we again see that $\mathfrak{G}$ is finitely generated, but with C_3 replaced by C_3/ϵ in (10). This gives the Mordell-Weil Theorem.

THEOREM 12.2.1. *If the ground field is a number field k, then $\mathfrak{G}$ is finitely generated.*

A fringe benefit of the height function concerns the torsion subgroup of $\mathfrak{G}$. Suppose that, for some $\mathfrak{A} \in \mathfrak{G}$, we have

$$H(\mathfrak{A}) > C_2^{1/3}. \qquad (12.2.17)$$

Then, by (iii) of (1), we must have

$$H(2\mathfrak{A}) \geqslant H(\mathfrak{A})^4/C_2 > H(\mathfrak{A}). \qquad (12.2.18)$$

Repeating this process gives

$$H(\mathfrak{A}) < H(2\mathfrak{A}) < H(4\mathfrak{A}) < \ldots \qquad (12.2.19)$$

so that $\mathfrak{A}, 2\mathfrak{A}, 4\mathfrak{A}, \ldots$ are all distinct, and so $\mathfrak{A}$ must not be a torsion element. It follows that

$$\mathfrak{G}_{\text{tors}} \subset \{\mathfrak{A} : H(\mathfrak{A}) \leqslant C_2^{1/3}\}. \qquad (12.2.20)$$

The constant C_2 is effectively computable and so, as promised in Chapter 8, this gives an (admittedly crude) effective procedure for computing $\mathfrak{G}_{\text{tors}}$.

3　　A computational improvement. It is apparent that the algorithm outlined in Sections 1, 2 gives a poor method in practice for deducing generators of $\mathfrak{G}$ from those for $\mathfrak{G}/2\mathfrak{G}$. This is primarily due to the poor value for the constant C_2 of property (iii) in (2.1). Recall that C_2 was obtained by first considering the duplication law $N(\mathbf{v}) = \big(N_j(\mathbf{v})\big)$, and then deriving a resultant matrix $R(\mathbf{v}) = \big(R_{ij}(\mathbf{v})\big)$, where each $R_{ij}(\mathbf{v})$ is a form in the indeterminates v_i, as described in equation (1.22). The constant C_2 was then expressed in terms of the coefficients of the polynomials R_{ij}. In the elliptic curve situation, this is viable since the computation of R_{ij} comes down to a pair of resultant computations on polynomials in two homogeneous variables $\binom{1}{x}$. Indeed this is done explicitly, for example, on p. 204 of Silverman (1986). In the genus 2 situation the computation of the R_{ij}, although effective in principle, is no longer viable. Each N_j is a form of degree 4 in four variables, and iterative resultant computations would be required to express a power of each variable as a member of the ideal generated by the N_j. The degrees of the R_{ij} in $\mathbf{v}$ would be enormous, as would be the resulting value of C_2.

It was pointed out in Flynn (1995a) that there is an alternative approach to the computation of C_2 using isogenies. Suppose that the ground field is a number field k. If necessary, consider the ground field k to be extended so that the curve can be transformed over k to the form

$$
\begin{aligned}
\mathcal{E}: \quad & Y^2 = X(X^2 + aX + b), \\
\mathcal{C}: \quad & Y^2 = F(X) = G_1(X)G_2(X)G_3(X),
\end{aligned}
\tag{12.3.1}
$$

where $a, b \in k$, and the G_i are quadratics defined over k as in (9.2.2). Let $\widehat{\mathcal{E}}$ and $\widehat{\mathcal{C}}$ be as in Chapter 10, Section 1. We shall also assume that the ground field k has been extended, if necessary, to include $\sqrt{b}$ in the elliptic curve case, and $\sqrt{b_1}, \sqrt{b_2}$ in the genus 2 case, where b_1, b_2 are as defined in (10.1.11). Recall that the isogeny from $\mathcal{E}$ to $\widehat{\mathcal{E}}$ of (10.1.7) takes the point (x, y) to $\big((x^2 + ax + b)/x, \, y - by/x^2\big)$, which induces a map on the x-coordinate. If we let $\mathbf{v}$ represent the projective x-coordinate $\binom{1}{x}$, then this map can be written in the form

$$
\begin{pmatrix} -1 & 1 \\ 2\sqrt{b} - a & 2\sqrt{b} + a \end{pmatrix} \tau \begin{pmatrix} -\sqrt{b} & 1 \\ \sqrt{b} & 1 \end{pmatrix} \mathbf{v},
\tag{12.3.2}
$$

where

$$
\tau: \quad (v_i) \longmapsto (v_i^2).
\tag{12.3.3}
$$

Note that the right hand matrix in (2) performs the role of diagonalizing the addition-by-$(0, 0)$ map on the projective x-coordinate. Composing with the dual isogeny gives the following way of factoring the pair of quartics $N(\mathbf{v})$ which give the duplication law on the projective x-coordinate.

$$N(\mathbf{v}) = M_1 \tau M_2 \tau M_3 \mathbf{v}, \qquad (12.3.4)$$

where

$$M_1 = \begin{pmatrix} 4 & 0 \\ 0 & 1 \end{pmatrix} \begin{pmatrix} -1 & 1 \\ 2\sqrt{\hat{b}} - \hat{a} & 2\sqrt{\hat{b}} + \hat{a} \end{pmatrix},$$

$$M_2 = \begin{pmatrix} -\sqrt{\hat{b}} & 1 \\ \sqrt{\hat{b}} & 1 \end{pmatrix} \begin{pmatrix} -1 & 1 \\ 2\sqrt{b} - a & 2\sqrt{b} + a \end{pmatrix}, \qquad (12.3.5)$$

$$M_3 = \begin{pmatrix} -\sqrt{b} & 1 \\ \sqrt{b} & 1 \end{pmatrix}.$$

In the genus 2 situation, using as usual the Kummer surface coordinates of (3.1.3) in $\mathbb{P}^3$, we can proceed in the same way. The duplication law is given by $N = (N_j) = (N_j(\mathbf{v}))$, where each N_j is a form of degree 4. There is a change of basis matrix M_3 which simultaneously diagonalizes addition by the points in the kernel of the isogeny, eventually leading to an identity of the same type as equation (4), but where M_1, M_2, M_3 are now 4×4 matrices whose entries involve $\sqrt{b_1}$ and $\sqrt{b_2}$. A complete description of M_1, M_2, M_3 is given in Flynn (1995a).

Once M_1, M_2, M_3 have been computed for a given curve over a number field k, they can be used to find the height constant C_2 without any need for resultants of polynomials. Clearly, for any $\mathbf{x} \in \mathbb{P}^n(k)$ and $M \in \mathbb{M}^n(k)$ we have

$$H(\tau(\mathbf{x})) = H(\mathbf{x})^2, \quad H(M \cdot \mathbf{x}) \leqslant H(M)H(\mathbf{x}). \qquad (12.3.6)$$

For any $\mathfrak{A} \in \mathfrak{G}$, consider the point $(\xi_i(\mathfrak{A})) \in \mathbb{P}^{2^g-1}(k)$. By repeated applications of (6) we see that

$$H_\kappa(2\mathfrak{A}) = H\left(N\left((\xi_i(\mathfrak{A}))\right)\right) = H\left(M_1 \tau M_2 \tau M_3 (\xi_i(\mathfrak{A}))\right)$$

$$\geqslant H(W_1^{-1})^{-1} H(W_2^{-1})^{-2} H(W_3^{-1})^{-4} H\left((\xi_i(\mathfrak{A}))\right)^4, \qquad (12.3.7)$$

and so

$$H_\kappa(2\mathfrak{A}) \geqslant H_\kappa(\mathfrak{A})^4 / C_2 \qquad (12.3.8)$$

where

$$C_2 = H(W_1^{-1}) H(W_2^{-1})^2 H(W_3^{-1})^4. \qquad (12.3.9)$$

This value of C_2 gives sufficient improvement in genus 2 that generators for $\mathfrak{G}$ can sometimes be computed when the coefficients are small. A detailed description is contained in Flynn (1995a), including the computation of several examples, such as the following, where we have already computed $\mathfrak{G}/2\mathfrak{G}$ in Chapter 11.

Example 12.3.1. *Let C_1 be the curve of Example 11.3.1; namely*

$$C_1: \quad Y^2 = F_1(X) = X(X-1)(X-2)(X-5)(X-6).$$

Then $\mathfrak{G}$ is generated by its 16 points of order 2 and $\{(3,6), \infty\}$.

Example 12.3.2. *Let*

$$C_2: \quad Y^2 = G_1(X)G_2(X)G_3(X) = (X^2+1)(X^2+2)(X^2+X+1),$$

with the curve of isogenous jacobian given by

$$\widehat{C_2}: \quad Y^2 = L_1(X)L_2(X)L_3(X) = (X^2-2X-2)(-X^2+1)(2X).$$

Then $\mathfrak{G}$ is generated by its four points of order 2 and $\{\infty^+, \infty^+\}$. Also, $\widehat{\mathfrak{G}}$ is generated by its eight points of order 2 and $\{(0,0), (-1/2, 3/4)\}$.

It should be conceded that the current techniques in the theory of heights are still far from satisfactory. The above trick with isogenies has made it possible to find generators for $\mathfrak{G}$ only when the coefficients of the curve of genus 2 are very small. Preferably it should be of the form C of equation (1) with G_1, G_2, G_3 defined over $\mathbb{Z}$, each with coefficients of absolute value less than 5. It would be nice to have genus 2 versions of the modern tricks currently applied to elliptic curves, such as those on pp. 55–58 of Cremona (1992).

Chapter 13

Rational points. Chabauty's Theorem

0   Introduction. A classical theorem due to Chabauty (1941a,b) states that a curve, defined over a number field k, has at most finitely many k-rational points when the jacobian has rank less than the genus of the curve. Curves of genus 2 provide the first source of nontrivial applications; we shall outline the main idea behind Chabauty's Theorem in this case. We shall show how the explicit structures on the jacobian variety outlined in previous chapters, such as the formal group law, can be used to give a bound on the number of rational points on a curve of genus 2 when the jacobian has rank 1. In some cases this bound is attained, allowing us to describe all rational points on the curve. Since the technique is local, it can be repeatedly applied to the same curve using different choices of prime p of good reduction; each choice of p gives a bound on the number of rational points on the curve over the original ground field, and so gives a new chance to improve the previous best bound.

1   Chabauty's Theorem. Let C be a curve of genus 2

$$C : \quad Y^2 = F(X) = f_0 + f_1 X + f_2 X^2 + f_3 X^3 + f_4 X^4 + f_5 X^5 + f_6 X^6, \quad (13.1.1)$$

where for simplicity we shall assume that the ground field is $\mathbb{Q}$. A more detailed account of the following ideas, including the generalization to number fields, can be found in Flynn (1995b). As usual, we shall further assume that X, Y have been adjusted by a constant (if necessary) so that all $f_i \in \mathbb{Z}$. Let $\mathfrak{G} = J(\mathbb{Q})$.

We first note that it is straightforward to find all Weierstrass points $(x, 0) \in C(\mathbb{Q})$, which only involves finding all roots of $F(X)$ in $\mathbb{Q}$. This only requires a finite search, since any such $x = a/b$, with $a, b \in \mathbb{Z}$ and $\gcd(a, b) = 1$, will satisfy $a \mid f_0$ and $b \mid f_6$. We can therefore concentrate on the non-Weierstrass points in $C(\mathbb{Q})$. Suppose that $P \in C(\mathbb{Q})$ is not a Weierstrass point. Then there is a correspondence

$$P \in C(\mathbb{Q}) \longleftrightarrow \{P, P\} \in \mathfrak{G}. \quad (13.1.2)$$

So it is sufficient to find all members of $\mathfrak{G}$ of the special form $\{P, P\}$. If $\mathfrak{G}$ has rank 0 then $\mathfrak{G} = \mathfrak{G}_{\text{tors}}$, which is finite, and so it is easy to find all such divisors, and therefore all of $C(\mathbb{Q})$. In Example 11.3.2, we can immediately deduce that $C(\mathbb{Q})$ consists only of the six Weierstrass points, including ∞.

The first nontrivial situation occurs when $\mathfrak{G}$ has rank 1, as in Examples 11.3.1 and 11.4.1. Let us say that we have found generators for $\mathfrak{G}$, with $\mathfrak{D}$ generating the infinite part, so that

$$\mathfrak{G} = \langle \mathfrak{G}_{\text{tors}}, \mathfrak{D} \rangle. \tag{13.1.3}$$

Let us now fix a prime p not dividing $2\,\mathrm{disc}\,(F)$, so that $\widetilde{C}$, the reduction of C mod p, is a curve of genus 2 over $\mathbb{F}_p$. Let $\widetilde{\mathfrak{D}}$ be the reduction of $\mathfrak{D}$ mod p, and let m be the torsion order of $\widetilde{\mathfrak{D}}$ on the jacobian over $\mathbb{F}_p$. Then the divisor

$$\mathfrak{E} = m \cdot \mathfrak{D} \in \mathfrak{G} \tag{13.1.4}$$

is in the kernel of reduction. That is to say, $\widetilde{\mathfrak{E}} = \widetilde{\mathfrak{O}}$. Clearly anything in $\mathfrak{G}$ can be written uniquely in the form

$$\mathfrak{A} + n \cdot \mathfrak{E}, \quad n \in \mathbb{Z}, \tag{13.1.5}$$

where

$$\mathfrak{A} \in \mathcal{U} = \{\mathfrak{B} + i \cdot \mathfrak{D} : \mathfrak{B} \in \mathfrak{G}_{\text{tors}} \text{ and } 1 \leqslant i \leqslant m - 1\}. \tag{13.1.6}$$

The set $\mathcal{U}$ is finite. Let $\mathfrak{A}$ be some fixed member of $\mathcal{U}$. We wish to bound the number of values of n such that the divisor in (5) is of the form $\{P, P\}$. First let $(z_i(\mathfrak{A}))$ and $(z_i(\mathfrak{E}))$ be the usual members of $\mathbb{P}^{15}(\mathbb{Q})$ as described at the beginning of Chapter 2, Section 2. Denote

$$\mathbf{a} = (a_i) = (z_i(\mathfrak{A})) \in \mathbb{P}^{15}(\mathbb{Q}), \tag{13.1.7}$$

where the a_i are chosen so that

$$a_i \in \mathbb{Z}, \quad \gcd(a_0, \ldots, a_{15}) = 1. \tag{13.1.8}$$

$\mathfrak{E}$ is in the kernel of reduction. Let $\mathbf{s}$ be its local parameter, as in (7.1.12). That is

$$\mathbf{s} = \mathbf{s}(\mathfrak{E}) = \begin{pmatrix} s_1(\mathfrak{E}) \\ s_2(\mathfrak{E}) \end{pmatrix} = \begin{pmatrix} z_1(\mathfrak{E})/z_0(\mathfrak{E}) \\ z_2(\mathfrak{E})/z_0(\mathfrak{E}) \end{pmatrix}. \tag{13.1.9}$$

We know, from Chapter 7, Section 1, that

$$|s_1|_p, |s_2|_p \leqslant p^{-1}. \tag{13.1.10}$$

The divisor $n \cdot \mathfrak{E}$ is also in the kernel of reduction, with local parameter given by

$$\mathbf{t} = \mathbf{t}(n) = \mathbf{s}(n \cdot \mathfrak{E}) = E(n \cdot L(\mathbf{s})), \tag{13.1.11}$$

where E and L are the formal exponential and logarithm power series described in Definition 7.2.2. This gives each of t_1, t_2 as power series in n defined over $\mathbb{Z}_p$. Let the power series σ_i, $i = 3, \ldots, 15$, be as in (7.1.11), extended to $i = 0, 1, 2$ as described immediately after (7.1.12). Then

$$\sigma(\mathbf{t}) = \big(\sigma_i(\mathbf{t})\big) \in \mathbb{P}^{15}(\mathbb{Z}_p[[n]]) \tag{13.1.12}$$

gives a complete description of $n \cdot \mathfrak{E} \in \mathfrak{G}$ as a vector of power series in n defined over $\mathbb{Z}_p$. Using the bilinear forms Φ_{ij}, defined over $\mathbb{Z}$, from Lemma 3.9.1, we see that

$$
\begin{aligned}
\big(\xi_i(\mathfrak{A} - n \cdot \mathfrak{E})\xi_j(\mathfrak{A} + n \cdot \mathfrak{E})\big) &= \Big(\Phi_{ij}\big(\mathbf{z}(\mathfrak{A}), \mathbf{z}(n \cdot \mathfrak{E})\big)\Big) \\
&= \Big(\Phi_{ij}\big(\mathbf{a}, \sigma(\mathbf{t})\big)\Big),
\end{aligned}
\tag{13.1.13}
$$

where the ξ_i are the Kummer surface coordinates defined in (3.1.3). In particular, provided that $\xi_4(\mathfrak{A} - n \cdot \mathfrak{E}) \neq 0$, we can select the fourth row to give the projective identity

$$\big(\xi_j(\mathfrak{A} + n \cdot \mathfrak{E})\big) = \Big(\Phi_{4j}\big(\mathbf{a}, \sigma(\mathbf{t})\big)\Big), \tag{13.1.14}$$

where each $\Phi_{4j}\big(\mathbf{a}, \sigma(\mathbf{t})\big)$ will again be a power series in n defined over $\mathbb{Z}_p$. If we now denote

$$\{(x, y), (u, v)\} = \mathfrak{A} + n \cdot \mathfrak{E}, \tag{13.1.15}$$

then, projectively,

$$
\begin{aligned}
(1, x + u, xu) &= \big(\xi_1(\mathfrak{A} + n \cdot \mathfrak{E}), \xi_2(\mathfrak{A} + n \cdot \mathfrak{E}), \xi_3(\mathfrak{A} + n \cdot \mathfrak{E})\big) \\
&= \Big(\Phi_{41}\big(\mathbf{a}, \sigma(\mathbf{t})\big), \Phi_{42}\big(\mathbf{a}, \sigma(\mathbf{t})\big), \Phi_{43}\big(\mathbf{a}, \sigma(\mathbf{t})\big)\Big).
\end{aligned}
\tag{13.1.16}
$$

We are interested in whether $\{(x, y), (u, v)\}$ is of the special form $\{P, P\}$. This will happen when

$$(x + u)^2 - 4xu = 0. \tag{13.1.17}$$

If we now define

$$
\begin{aligned}
\theta(n) &= \theta_{\mathfrak{A}, \mathfrak{E}}(n) \\
&= \Big(\Phi_{42}\big(\mathbf{a}, \sigma(\mathbf{t})\big)\Big)^2 - 4\Phi_{41}\big(\mathbf{a}, \sigma(\mathbf{t})\big) \cdot \Phi_{43}\big(\mathbf{a}, \sigma(\mathbf{t})\big) \\
&\in \mathbb{Z}_p[[n]],
\end{aligned}
\tag{13.1.18}
$$

then we have the following implication.

$$\text{If } \mathfrak{A} + n \cdot \mathfrak{E} \text{ is of the form } \{P, P\} \text{ then } \theta(n) = 0. \qquad (13.1.19)$$

Note that this remains true even in the case we have so far excluded, namely $\xi_4(\mathfrak{A} - n \cdot \mathfrak{E}) = 0$, when all of the $\Phi_{4j}(\mathbf{a}, \sigma(\mathbf{t}))$ are zero, and so again $\theta(n) = 0$. This case may give extra unwanted solutions, where $\xi_4(\mathfrak{A} - n \cdot \mathfrak{E}) = 0$ and so $\theta(n) = 0$, even though $\mathfrak{A} + n \cdot \mathfrak{E}$ is not of the form $\{P, P\}$ [and so the implication (19) is not reversible, even in $\mathbb{Z}_p$]. This does not seem to cause problems in practice; one can either try to recognize these unwanted solutions explicitly and set them aside, or there is the option of recomputing a variation of $\theta(n)$ using a row of the Φ_{ij} different from the fourth; that is, replacing 4 with any of 1, 2 or 3 in (14) [but beware that each choice of row may lead to a $\theta(n)$ with extra unwanted solutions]. Write

$$\theta(n) = c_0 + c_1 n + c_2 n^2 + \dots, \quad c_i \in \mathbb{Z}_p. \qquad (13.1.20)$$

We can make a much stronger statement about the coefficients c_i than merely that they are in $\mathbb{Z}_p$. We first recall that (7.2.6) describes what denominators can occur in the power series L_1, L_2, E_1, E_2; namely at worst a denominator of $i!j!$ can occur in the coefficient of $s_1^i s_2^j$. Since $i!j! | (i+j)!$, and since $|s_1|_p, |s_2|_p \leqslant p^{-1}$, we can (as in Chapter 7, Section 5) use the standard estimate

$$|r!| > p^{-r/(p-1)}. \qquad (13.1.21)$$

This first gives that $|L_i(\mathbf{s})|_p \leqslant p^{-1}$ for $i = 1, 2$. It also follows that each power series $\mathbf{t}(n) = (t_i(n))$ of equation (11) has the property that the coefficient of n^j converges to 0 in $\mathbb{Z}_p$ as $j \to \infty$. Since all of the σ_i and Φ_{ij} are defined over $\mathbb{Z}$, the same property must be true of the coefficients c_j of $\theta(n)$ in equation (20). That is

$$|c_j|_p \longrightarrow 0, \text{ as } j \to \infty. \qquad (13.1.22)$$

The following standard theorem is proved on p. 62 of Cassels (1986).

THEOREM 13.1.1 (Strassman). *Let* $\theta(X) = c_0 + c_1 X + \dots \in \mathbb{Z}_p[[X]]$ *satisfy* $c_j \to 0$ *in* $\mathbb{Z}_p$. *Define* ℓ *uniquely by:* $|c_\ell|_v \geqslant |c_j|_v$ *for all* $j \geqslant 0$, *and* $|c_\ell|_v > |c_j|_v$ *for all* $j > \ell$. *Then there are at most* ℓ *values of* $x \in \mathbb{Z}_p$ *such that* $\theta(x) = 0$ *and* $|x|_v \leqslant 1$.

It is now possible to give a description of the strategy for bounding the cardinality of $\mathcal{C}(\mathbb{Q})$. For each $\mathfrak{A}$ in the finite set $\mathcal{U}$ of equation (6), we compute the coefficients c_j of $\theta_{\mathfrak{A}, \mathfrak{E}}(n)$ to sufficient p-adic accuracy to determine the coefficient c_ℓ which satisfies the condition of Strassman's Theorem.

Then the number of $n \in \mathbb{Z}_p$, and hence the number of $n \in \mathbb{Z}$, satisfying $\theta_{\mathfrak{A},\mathfrak{E}}(n) = 0$ will be bounded by $\ell = \ell(\mathfrak{A})$. This gives a bound on the number of n such that the divisor $\mathfrak{A} + n \cdot \mathfrak{E}$ is of the form $\{P, P\}$ for some $P \in \mathcal{C}(\mathbb{Q})$. The bound differs from the truth in two ways: (i) by the number of extra unwanted solutions of the type described just after (19), which we can try to deal with, and (ii) by the number of n in $\mathbb{Z}_p$, but not in $\mathbb{Z}$, such that $\mathfrak{A} + n \cdot \mathfrak{E}$ is of the form $\{P, P\}$.

The sum of these bounds $\ell = \ell(\mathfrak{A})$ over all $\mathfrak{A} \in \mathcal{U}$ gives a bound on the total number of possible divisors of the form $\{P, P\}$ in $\mathfrak{G}$, and so gives a bound on the number of non-Weierstrass points on $\mathcal{C}(\mathbb{Q})$, as required. If the bound is not the same as the number of known non-Weierstrass points on $\mathcal{C}(\mathbb{Q})$, then we can repeat the above process with a new prime, which gives a new chance of determining $\mathcal{C}(\mathbb{Q})$ completely. When it is suspected that there is a missing member of $\mathcal{C}(\mathbb{Q})$ still to be found, then the local information can also be used to speed up the search. Despite everything that can in principle go wrong, we have so far found the results to be highly promising, with the bound on $\#\mathcal{C}(\mathbb{Q})$ typically equalling, or coming within 1 of, the number of known points on $\mathcal{C}(\mathbb{Q})$.

The equipment needed to implement the above process is all available, namely the equations for the bilinear forms Φ_{ij} (available by anonymous ftp, as described in Appendix II) and the ability to derive as many terms as desired of the power series σ_i, L_i, E_i above. For the power series, one can use the methods described in Chapter 7. For the lazy reader, we have also provided by anonymous ftp all of these power series, as well as the $\mathcal{F}_i$ for the formal group, computed for a general curve of genus 2 up to all terms of degree 7 in $\mathbf{s}$. We have found in practice that these are more than sufficient for typical applications. Indeed, it seems nearly always to be enough to know the L_i, E_i up to terms of degree 3 in $\mathbf{s}$.

The above discussion is a special case of ideas which apply in genus > 1 over a number field. The following was proved by Chabauty (1941a,b).

THEOREM 13.1.2. *Let $\mathcal{C}$ be a curve of genus $g > 1$ defined over a number field k. If the jacobian of $\mathcal{C}$ has rank less than g, then $\mathcal{C}(k)$ is finite.*

This result is of course superceded by Faltings' work, which gives the same conclusion with no condition on the rank. However, when applicable, the methods used by Chabauty give a much better bound on the cardinality of $\mathcal{C}(k)$. It is rather a curiosity of the literature that Chabauty's result has existed for so long, but is only now becoming useful as a flexible tool for solving Diophantine problems.

2   A worked example. We shall illustrate the ideas of Section 1 with the following example, in which $C(\mathbb{Q})$ is found completely.

Example 13.2.1. Let $\widehat{C}_2$ be the curve of Example 12.3.2, namely

$$\widehat{C}_2: \quad Y^2 = L_1(X)L_2(X)L_3(X) = (X^2 - 2X - 2)(-X^2 + 1)(2X). \quad (13.2.1)$$

Then $\widehat{C}_2(\mathbb{Q}) = \{(0,0), \infty, (\pm 1, 0), (-1/2, \pm 3/4)\}$.

Proof. The quadratic $L_1(X) = X^2 - 2X - 2$ is irreducible over $\mathbb{Q}$, and so the only Weierstrass points in $C(\mathbb{Q})$ are $(0,0), \infty$ and $(\pm 1, 0)$. It remains to show that $(-1/2, \pm 3/4)$ are the only non-Weierstrass points. It has already been shown in Example 12.3.2 that $\mathfrak{G}$ is generated by the eight elements of order 2 and the infinite generator $\{(0,0), (-1/2, 3/4)\}$. In fact, it will be more convenient to adjust this by adding the 2-torsion element $\{\infty, (-1,0)\}$, and use instead

$$\begin{aligned} \mathfrak{D} &= \{(0,0), (-1/2, 3/4)\} + \{\infty, (-1,0)\} \\ &= \{(-\frac{1}{2} + \frac{1}{2}\sqrt{-15}, -12), (-\frac{1}{2} - \frac{1}{2}\sqrt{-15}, -12)\}. \end{aligned} \quad (13.2.2)$$

Then

$$\mathfrak{G}_{\text{tors}} = \langle \{(-1,0), (1,0)\}, \{\infty, (0,0)\}, \{\infty, (-1,0)\} \rangle \quad (13.2.3)$$

and

$$\mathfrak{G} = \langle \mathfrak{G}_{\text{tors}}, \mathfrak{D} \rangle. \quad (13.2.4)$$

The discriminant of $L_1(X)L_2(X)L_3(X)$ is $2^{14}3^3$, and so the reduction of $\widehat{C}_2$ over $\mathbb{F}_p$ is a curve of genus 2 for any prime $p \neq 2, 3$. We shall choose $p = 5$, and let $\widetilde{\mathfrak{D}}$ denote the reduction of $\mathfrak{D}$ mod 5. The jacobian over $\mathbb{F}_5$ has 24 elements; the eight elements of order 2 in $\mathfrak{G}$ inject under the reduction map, and $\widetilde{\mathfrak{D}}$ is of order 3.¶ The divisor

$$\mathfrak{E} = 3 \cdot \mathfrak{D} = \{Q, Q'\},$$
$$\text{where } Q = (\frac{-197}{10} + \frac{1}{10}\sqrt{34185}, \frac{32652}{5} - \frac{4416}{125}\sqrt{34185}), \quad (13.2.5)$$

is in the kernel of reduction, and anything in $\mathfrak{G}$ can be written in the form

$$\mathfrak{A} + n \cdot \mathfrak{E}, \quad n \in \mathbb{Z}, \quad (13.2.6)$$

¶ This is why for simplicity we chose $\mathfrak{D}$ instead of $\{(0,0), (-1/2, 3/4)\}$, since the latter reduces to an element of order 6 over $\mathbb{F}_5$.

where $\mathfrak{A}$ is in the finite set given in (1.6) with $m = 3$. However, for simplicity, it is convenient to use instead

$$\mathfrak{A} \in \mathcal{U} = \{\mathfrak{B} + i \cdot \mathfrak{D} : \mathfrak{B} \in \widehat{\mathfrak{S}}_{\text{tors}} \text{ and } -1 \leqslant i \leqslant 1\}. \tag{13.2.7}$$

Note that this is legitimate, since this definition of $\mathcal{U}$ is equivalent to that in (1.6) modulo multiples of $\mathfrak{E}$. For each $\mathfrak{A} \in \mathfrak{U}$, we wish to bound the number of values of n such that the divisor in (6) is of the form $\{P, P\}$. We first observe that, under the reduction map,

$$\mathfrak{A} + n \cdot \mathfrak{E} \longmapsto \widetilde{\mathfrak{A}}. \tag{13.2.8}$$

If $\mathfrak{A} + n \cdot \mathfrak{E}$ were of the form $\{P, P\}$ then $\widetilde{\mathfrak{A}}$ would have to be of the form $\{\widetilde{P}, \widetilde{P}\}$ over $\mathbb{F}_5$. So we need only consider those $\mathfrak{A} \in \mathcal{U}$ such that $\widetilde{\mathfrak{A}}$ is of this form. Of the 24 members of $\mathcal{U}$, there are only three possible choices of $\mathfrak{A}$ for which this is true.

Case 1. $\mathfrak{A} = \{(-1/2, 3/4), (-1/2, 3/4)\}.$
Case 2. $\mathfrak{A} = \{(-1/2, -3/4), (-1/2, -3/4)\}.$ $\qquad$ (13.2.9)
Case 3. $\mathfrak{A} = \mathfrak{D}.$

We first analyse $E(n \cdot L(\mathbf{s}))$, since this will be required in all 3 cases. First note that

$$\mathbf{s} = \mathbf{s}(\mathfrak{E}) = \begin{pmatrix} s_1(\mathfrak{E}) \\ s_2(\mathfrak{E}) \end{pmatrix} = \begin{pmatrix} z_1(\mathfrak{E})/z_0(\mathfrak{E}) \\ z_2(\mathfrak{E})/z_0(\mathfrak{E}) \end{pmatrix}$$
$$= \begin{pmatrix} 109\,944\,355/196\,684\,227 \\ -30\,884\,675/393\,368\,454 \end{pmatrix} \tag{13.2.10}$$

and so

$$|s_1|_5 = 5^{-1}, \quad |s_2|_5 = 5^{-2}. \tag{13.2.11}$$

It will be sufficient to find $E(n \cdot L(\mathbf{s}))$ modulo 5^4. On substituting $s_1 \equiv 490$, and $s_2 \equiv 550 \pmod{5^4}$ into L_1, L_2 of (7.2.7), we get

$$L(\mathbf{s}) = \begin{pmatrix} L_1(\mathbf{s}) \\ L_2(\mathbf{s}) \end{pmatrix} \equiv \begin{pmatrix} 240 \\ 175 \end{pmatrix} \pmod{5^4}. \tag{13.2.12}$$

Note that, even taking the denominators $1/i!j!$ into account, we can ignore terms of degree $\geqslant 4$ in $\mathbf{s}$ when working modulo 5^4, and so the first few terms given in (7.2.7) are sufficient. Similarly, we can now use E_1, E_2 of (7.2.7) to get

$$E(n \cdot L(\mathbf{s})) = \begin{pmatrix} E_1(n \cdot L(\mathbf{s})) \\ E_2(n \cdot L(\mathbf{s})) \end{pmatrix} \equiv \begin{pmatrix} 240n + 250n^3 \\ 175n + 375n^3 \end{pmatrix} \pmod{5^4}. \tag{13.2.13}$$

This can be summarized by saying that if $\mathbf{t}$ is the local parameter for $n \cdot \mathcal{E}$, then each of t_1, t_2 can be expressed as power series in n over $\mathbb{Z}_5$, with

$$t_1 \equiv 240n + 250n^3, \quad t_2 \equiv 175n + 375n^3 \quad (\text{mod } 5^4). \qquad (13.2.14)$$

We now compute the power series arising from each of the three possible choices for $\mathfrak{A}$.

Case 1. $\mathfrak{A} = \{(-1/2, 3/4), (-1/2, 3/4)\}$. Following Section 1, we first find the integers $a_0, \ldots, a_{15}$ described in (1.7),(1.8) such that $(a_i) = (z_i(\mathfrak{A}))$. Substituting these into

$$\left(\Phi_{42}(\mathbf{a}, \sigma(\mathbf{t}))\right)^2 - 4\Phi_{41}(\mathbf{a}, \sigma(\mathbf{t})) \cdot \Phi_{43}(\mathbf{a}, \sigma(\mathbf{t})), \qquad (13.2.15)$$

we obtain a power series in t_1, t_2 defined over $\mathbb{Z}$ which equals 0 when $\mathbf{a} + \sigma(\mathbf{t})$ corresponds to a divisor of the form $\{P, P\}$. Ignoring terms of degree > 3 in $\mathbf{t}$, and reducing the coefficients modulo 5^4, we find that this power series is congruent (mod 5^4) to

$$\begin{aligned}
349t_1 + 487t_2 + 615t_1^2 + 261t_1 t_2 + 549t_2^2 \\
+ 218t_1^3 + 221t_1^2 t_2 + 557t_1 t_2^2 + 401t_2^3 + \ldots
\end{aligned} \qquad (13.2.16)$$

On substituting (14) into (16), we obtain the power series $\theta_{\mathfrak{A}, \mathfrak{E}}(n) \in \mathbb{Z}_5[[n]]$ of (1.18) modulo 5^4 as

$$\theta_{\mathfrak{A}, \mathfrak{E}}(n) \equiv 235n + 375n^2 \quad (\text{mod } 5^4). \qquad (13.2.17)$$

But $|235|_5 = 5^{-1}$, $|375|_5 = 5^{-3}$ and $|c_j| < 5^{-4}$ for any coefficient c_j of n^j with $j > 2$. It follows from Strassman's Theorem 1.1 that there is at most one value of $n \in \mathbb{Z}_5$ at which $\theta_{\mathfrak{A}, \mathfrak{E}}(n) = 0$, namely $n = 0$. Therefore, $n = 0$ is the only $n \in \mathbb{Z}$ such that $\mathfrak{A} + n \cdot \mathfrak{E}$ is of the form $\{P, P\}$.

Case 2. $\mathfrak{A} = \{(-1/2, -3/4), (-1/2, -3/4)\}$. This choice of $\mathfrak{A}$ is merely the negative of that in case 1. We observe that

$$\begin{aligned}
&\{(-1/2, 3/4), (-1/2, 3/4)\} + n \cdot \mathfrak{E} \\
&= \{(x, y), (x, y)\} \\
\Longleftrightarrow\ &\{(-1/2, -3/4), (-1/2, -3/4)\} + (-n) \cdot \mathfrak{E} \\
&= \{(x, -y), (x, -y)\},
\end{aligned} \qquad (13.2.18)$$

and so it follows immediately from case 1 that again $n = 0$ is the only possibility. Note that, if we were to follow through the same steps as in Case 1, the resulting $\theta(n)$ would be as in (17), but with n replaced by $-n$.

Case 3. $\mathfrak{A} = \mathfrak{D}$, so that $a_0 = 1$ and $a_i = 0$ for $i > 0$. Then (15) has only terms of even degree in **t** and the lowest degree terms have degree 6 in **t**. These terms are

$$4(-2t_1^5 t_2 + 4t_1^4 t_2^2 + 6t_1^3 t_2^3 - 4t_1^2 t_2^4 - 4t_1 t_2^5) + \text{terms of degree} \geqslant 8. \quad (13.2.19)$$

On substituting (14) into (19), we obtain the power series $\theta_{\mathfrak{A},\mathfrak{C}}(n) \in \mathbb{Z}_5[[n]]$ of (1.18) modulo 5^8 as

$$\theta_{\mathfrak{A},\mathfrak{C}}(n) \equiv 156250n^6 \pmod{5^8}. \quad (13.2.20)$$

But $|156250|_5 = 5^{-7}$, which is strictly greater than $|c_j|_5$ for any other co-efficient c_j. Therefore, after taking out the factor of n^6, we can again use Strassman's Theorem to see that $n = 0$ is the only solution.

We have now found all possible divisors of the form $\{P, P\}$ in $\mathfrak{G}$, namely $\{(-1/2, 3/4), (-1/2, 3/4)\}$ and $(-1/2, -3/4), (-1/2, -3/4)\}$, from cases 1 and 2, respectively. It follows that $((-1/2, \pm 3/4)$ are the only two possible non-Weierstrass points on $\widehat{\mathcal{C}}_2$, as required.

We can summarize this method of attack for finding $\mathcal{C}(\mathbb{Q})$ as follows.

Step 1. Try to find $\mathfrak{G}_{\text{tors}}$ and $\mathfrak{G}/2\mathfrak{G}$, and show that $\mathfrak{G}$ has rank 1.
Step 2. Use heights to find a generator $\mathfrak{D}$ such that $\mathfrak{G} = \langle \mathfrak{G}_{\text{tors}}, \mathfrak{D} \rangle$.
Step 3. Find the Chabauty bound with respect to some good prime p.

In applying these three steps to the above curve $\widehat{\mathcal{C}}_2$, by far the majority of the time was used by the height computation of Step 2. For other examples, with larger coefficients, this height computation will become prohibitive. If one is interested merely in the final result of Step 3 bounding the cardinality of $\mathcal{C}(\mathbb{Q})$, then it is natural to ask whether Step 2 can be bypassed. Consider the curve $\widehat{\mathcal{C}}_2$ of the above example, and suppose that we have found $\widehat{\mathfrak{G}}_{\text{tors}}$ and $\mathfrak{G}/2\widehat{\mathfrak{G}}$. Let $\mathfrak{D}$ be as in equation (2). Since $\mathfrak{G}_{\text{tors}}$ contains only 2-torsion, we can at this point say that there exists an integer w and $\mathfrak{D}_0 \in \widehat{\mathfrak{G}}$ such that

$$\mathfrak{D} = w \cdot \mathfrak{D}_0, \text{ and } \widehat{\mathfrak{G}} = \langle \widehat{\mathfrak{G}}_{\text{tors}}, \mathfrak{D}_0 \rangle. \quad (13.2.21)$$

Also define

$$\mathfrak{E}_0 = 3 \cdot \mathfrak{D}_0 \quad (13.2.22)$$

so that $\mathfrak{E}_0 = w \cdot \mathfrak{E}$, where $\mathfrak{E}$ is as in (5). The height computation of Step 2 is required to be sure that we can take $w = 1$ and $\mathfrak{D} = \mathfrak{D}_0$, but some weaker information can quickly be obtained about w. First of all, from the descent via isogeny performed in Example 11.4.1, we know that $\mathfrak{D} \notin 2\widehat{\mathfrak{G}}$ and so w is odd. We also know that $\widetilde{\mathfrak{D}} = w \cdot \widetilde{\mathfrak{D}}_0$ in the reduction of $\widehat{\mathfrak{G}}$ modulo 5, and it is a finite computation to check that this implies

$$\mathfrak{E}_0 \in \text{kernel of reduction (mod 5)}. \quad (13.2.23)$$

It is also a finite computation (mod 5) to check that $\mathfrak{D} \notin 3\widehat{\mathfrak{G}}$, and a similar computation (mod 11) that $\mathfrak{D} \notin 5\widehat{\mathfrak{G}}$. Thus

$$3 \nmid w, \quad 5 \nmid w. \tag{13.2.24}$$

Note that, since $5 \nmid w$, there is no w-torsion in the kernel of reduction mod 5, which has a well defined map $1/w$. In particular, $\mathfrak{E}_0 = (1/w) \cdot \mathfrak{E}$. The fact that $3 \nmid w$ (and the consequent existence of λ, μ such that $3\lambda + w\mu = 1$) gives that any member of $\widehat{\mathfrak{G}}$ can be written in the form

$$\mathfrak{A} + (v/w) \cdot \mathfrak{E}, \quad v \in \mathbb{Z}, \tag{13.2.25}$$

where $\mathfrak{A} \in \mathcal{U}$, the same finite set as defined in (1.6). But the proof given in the above example found, for each $\mathfrak{A}$, all $n \in \mathbb{Z}_3$ such that $\mathfrak{A} + n \cdot \mathfrak{E}$ is of the form $\{P, P\}$, and so this includes all possible v/w with $5 \nmid w$. This means that the final conclusion about $\widehat{C}_2(\mathbb{Q})$ can be made independently of the height computation in Step 2.

For a general curve C of genus 2, suppose that we have found $\mathfrak{G}_{\text{tors}}$ and $\mathfrak{G}/2\mathfrak{G}$ with $\mathfrak{G}$ of rank 1 and $\mathfrak{D} \in \mathfrak{G}$ of infinite order, but that we have not performed Step 2. Suppose that p is a prime of good reduction. It is a finite process to determine whether there is a torsion point $\mathfrak{A}_1$ such that $\mathfrak{D} = p \cdot \mathfrak{D}_1 + \mathfrak{A}_1$ for some $\mathfrak{D}_1$. If there is, then repeat the process, but applied to $\mathfrak{D}_1$ instead of $\mathfrak{D}$. After a finite number of steps, we find a divisor $\mathfrak{D}_n$ such that $\mathfrak{D}_n + \mathfrak{A}$ is not in $p \cdot \mathfrak{G}$ for any $\mathfrak{A} \in \mathfrak{G}_{\text{tors}}$ (we hope to prove this by a finite field argument). It follows that every $\mathfrak{B} \in \mathfrak{G}$ is of the form $\mathfrak{B} = w_1 \cdot \mathfrak{D}_n + \mathfrak{A}_1$, where $\mathfrak{A}_1 \in \mathfrak{G}_{\text{tors}}$ and $w_1 \in \mathbb{Z}_p$. Now one can again proceed with an argument in $\mathbb{Z}_p$ rather than in $\mathbb{Z}$, and bypass the time-consuming Step 2.

We also draw the reader's attention to the paper of Coleman (1985a), which includes the following two results.

Proposition 13.2.1. *Let C be a curve of genus 2 defined over $\mathbb{Q}$, and $p \geqslant 5$ be a prime of good reduction. If the jacobian of C has rank at most 1 and $\widetilde{C}$ is the reduction of C mod p then $\#C(\mathbb{Q}) \leqslant \#\widetilde{C}(\mathbb{F}_p) + 2$.*

Proposition 13.2.2. *Let C be the curve of genus 2,*

$$C: \quad Y^2 = X(X^2 - 1)(X - 1/\lambda)(X^2 + aX + b), \tag{13.2.26}$$

with $\lambda, a, b \in \mathbb{Z}$. Suppose $3^{2r} | \lambda$, for some $r > 0$, and 3 does not divide $b(1 - a + b)(1 + a + b)$, and that the jacobian of C has rank at most 1. Then

$\mathcal{C}(\mathbb{Q})$ *contains precisely the points* $(0,0),(1,0),(-1,0),(1/\lambda,0)$ *and the two rational points at infinity.*

We already have an example for which Proposition 1 determines $\mathcal{C}(\mathbb{Q})$ completely, namely the curve $\mathcal{C}_1$ of Example 11.3.1. In this case, we have $\#\mathcal{C}(\mathbb{F}_7) = 8$, and $\mathcal{C}_1(\mathbb{Q})$ contains the known six rational Weierstrass points, together with $(3, \pm 6)$ and $(10, \pm 120)$. By Proposition 1, these must give all of $\mathcal{C}_1(\mathbb{Q})$. This piece of good luck seems unlikely to be common, and in general we cannot expect Proposition 1 to determine $\mathcal{C}(\mathbb{Q})$ completely. For the curve $\widehat{\mathcal{C}}_2$ in our worked example, $\#\widehat{\mathcal{C}}_2(\mathbb{F}_5) = 6$, which only bounds $\widehat{\mathcal{C}}_2$ by 8, with larger primes giving a worse bound, and so the more detailed analysis given in the worked example was indeed required.

Proposition 2 gives a more likely source of examples for which $\mathcal{C}(\mathbb{Q})$ can be found competely. It was pointed out in Flynn (1995c) that the jacobian of the curve

$$Y^2 = X(X^2 - 1)(X + \frac{1}{9})(X^2 - 4X - 1) \qquad (13.2.27)$$

has rank 1 over $\mathbb{Q}$ and so, by Proposition 2, there are no $\mathbb{Q}$-rational points on the curve apart from $(0,0),(1,0),(-1,0),(-1/9,0)$ and the two rational points at infinity. The derivation of the rank can be performed using descent via isogeny, in a similar style to Example 11.4.1.

Chabauty's Theorem has also been used to obtain conditional bounds on the number of rational points on the Fermat curves; see McCallum (1992).¶

The current genus 2 record for the largest number of known rational points is held by the curve [found by Keller and Kulesz (1995)]

$$Y^2 = 278\,271\,081X^2(X^2 - 9)^2 - 229\,833\,600(X^2 - 1)^2, \qquad (13.2.28)$$

which has at least 588 rational points. It is not known whether there are any other rational points. The curve's group of automorphisms is of order 12, generated by $(X, Y) \mapsto (-X, Y)$, $(X, Y) \mapsto ((X + 3)/(X - 1), Y/(X - 1)^3)$ and the hyperelliptic involution $(X, Y) \mapsto (X, -Y)$, so it is sufficient to give 49 representative points modulo the group of automorphisms [there is a list on p. 1469 of Keller and Kulesz (1995)]. As will be apparent in the next chapter, the jacobian is isogenous to the product of two elliptic curves.

¶ Pre-Wiles! In McCallum (1988), the jacobian of certain quotients of the Fermat curves are shown to have systematically occurring non-trivial members of the Shafarevich-Tate group.

Chapter 14

Reducible jacobians

0   Introduction. It can happen that an abelian variety of dimension 2 is isogenous to a product of two (not necessarily distinct) abelian varieties of dimension 1, i.e. elliptic curves. The first example¶ of this seems to have been given by Legendre at the age of 80 in the *troisième supplément* to his *Théorie des fonctions elliptiques*: in a review Jacobi (1832) gave one form of Theorem 1.1 below. There are infinitely many cases indexed by a natural number as parameter, but Jacobi's is particularly straightforward. We have already met it in Chapter 9, where it had to be excluded from the discussion. We shall treat it in this chapter. It occurs quite frequently for naturally arising curves.† The other cases of reducibility appear to be much more difficult to handle, and we do not make the attempt. There is a treatment in Frey & Kani (1991) and a rather inconclusive discussion of the construction of explicit examples in Frey (1995). For a different approach and further references, see Kuhn (1988). See also Ruppert (1990) [use of the complex structure], Grant (1994c) [use of L-functions], Stoll (1995), pp. 1343–1344 [use of L-functions], and Flynn, Poonen & Schaefer (1995) [simple proof of irreducibility in a special case with probable wide applicability].

1   The straightforward case. Since what we are doing in this section is geometry, we work in the algebraic closure. We say that two curves $Y^2 = F(X)$ are equivalent if they are taken into one another by a fractional linear transformation of X and the related transformation of Y [see (1.1.3)].

¶ For the history and background, see Krazer (1903), Kap. 11, especially Section 2 and Hudson (1905), Chapter XVIII. Hudson calls the corresponding Kummer *elliptic*. In this chapter he also discusses complex and real multiplication, which we leave until Chapter 15.

† The jacobians of many of the curves listed by Merriman & Smart (1993) are reducible.

THEOREM 14.1.1. *The following properties of a curve C of genus 2 are equivalent:*

(i) *It is equivalent to a curve*

$$Y^2 = c_3 X^6 + c_2 X^4 + c_1 X^2 + c_0 \qquad (14.1.1)$$

with no terms of odd degree in X.

(ii) *It is equivalent to a curve*

$$Y^2 = G_1(X)G_2(X)G_3(X), \qquad (14.1.2)$$

where the quadratics $G_j(X)$ are linearly dependent.

(iii) *C is equivalent to*

$$Y^2 = X(X-1)(X-a)(X-b)(X-ab) \qquad (14.1.3)$$

for some a, b.

(iv) *A Göpel tetrahedron on the Kummer has its four vertices coplanar.*

If one (and so all) of these conditions is satisfied, the jacobian of C is reducible.

There are two maps of (1) into elliptic curves

$$\mathcal{E}_1 : Y^2 = c_3 Z^3 + c_2 Z^2 + c_1 Z + c_0, \qquad (14.1.4)$$

with $Z = X^2$ and

$$\mathcal{E}_2 : V^2 = c_0 U^3 + c_1 U^2 + c_2 U + c_3, \qquad (14.1.5)$$

with $U = X^{-2}$, $V = Y X^{-3}$. These extend to maps of the jacobian, which is therefore reducible.¶

¶ By general functoriality arguments. Alternatively, let $\phi : C \to \mathcal{E}$ be a map of a curve of genus 2 into an elliptic curve. The extension to the jacobian $J(C)$ maps a divisor $\mathfrak{X} = \{(x, y), (u, v)\}$ into $\phi(x, y) + \phi(u, v)$, where $+$ is the group law on $\mathcal{E}$. Suppose that C is given by (1) and that $\mathfrak{X}$ is in the kernel of the maps into both $\mathcal{E}_j$. Then $x^2 = u^2$, $y = -v$ from $\mathcal{E}_1$ and $x^{-2} = u^{-2}$, $yx^{-3} = -vu^{-3}$ from $\mathcal{E}_2$. Hence $x = u$ and $\mathfrak{X}$ is in the canonical class $\mathfrak{O}$. It follows that $J(C)$ is isogenous to $\mathcal{E}_1 \times \mathcal{E}_2$.

The form (1) is a special case of (2). Conversely, suppose that (2) holds. No two of the G_j have a common root, so there are two distinct singular quadratics in the pencil, say $(X - \lambda)^2$ and $(X - \mu)^2$. We map into a form of type (1) by taking $(X - \lambda)/(X - \mu)$ for X.

The form (3) is a special case of (2) with $G_1 = X$, $G_2 = (X-1)(X-ab)$, and $G_3 = (X - a)(X - b)$. Conversely (2) is reduced to (3) by a fractional linear transformation which takes the two roots of G_1 to 0 and ∞ and one of the roots of G_2 to 1.

Any Göpel tetrahedron can be taken, by adding the divisor belonging to one of its nodes, into a Göpel tetrahedron with one of its vertices at $\mathbf{o} = (0, 0, 0, 1)$. The other vertices now correspond to G_j in a factorization (2) [cf. Chapter 9]. If $G_j = g_{j2}X^2 + g_{j1} + g_{j0}$, the coordinates of the corresponding node are $(g_{j2}, g_{j1}, g_{j0}, *)$, where the value of $*$ is immaterial. The equivalence of (ii) and (iv) is now immediate. This concludes the proof.

Criterion (iii) is Jacobi's. It is particularly easy to apply mechanically.

Kummers which satisfy Criterion (iv) were investigated by Cayley (1846) [cf. Hudson (1905), Section 56, pp. 89–90]; so long before Kummer invented his eponymous surfaces. He called them *tetrahedroids* because a tetrahedron is the main element of his construction. A metrical version is even earlier: the *wave surface* introduced by Fresnel in his 1821 memoir on double refraction. [cf. Hudson (1905) Chapter X, pp. 100–112].

2   An awful warning. We now revert to a not necessarily algebraically closed ground field. If a curve comes in the shape (1.2) with the G_j defined over the ground field, that is not necessarily the correct factorization for the application of the criterion. Consider¶

$$
\begin{aligned}
G_1 &= 9X^2 - 28X + 18, \\
G_2 &= X^2 + 12X + 2, \\
G_3 &= X^2 - 2.
\end{aligned}
\tag{14.2.1}
$$

These are linearly independent. Both G_1 and G_2 split over $\mathbb{Q}(\sqrt{34})$. If we combine the factors of G_1 and G_2 differently,† the criterion applies, and so

¶ This is also the case with Legendre's curve $Y^2 = X(1 - X^2)(1 - \kappa^2 X^2)$. The curve considered here arose in a so far unsuccessful attack on a challenge problem of Serre, to which we revert in Chapter 18. We were blissfully ignorant of the possible reducibility of the jacobian of this curve until forced to that conclusion by its unexpected and anomalous behaviour.

† Curiously, it does not matter which way.

the jacobian is reducible. More precisely $9G_1 = LL'$, $G_2 = MM'$ where $L = 9X - 14 - \sqrt{34}$, $M = X + 6 - \sqrt{34}$ and $'$ denotes the conjugate. Then LM, $L'M'$ and G_3 are linearly dependent.

3   The reverse process. Given two elliptic curves, there is a curve of genus 2 whose jacobian is isogenous to their product. If the elliptic curves and their Weierstrass points are rational and we permit a 'twist' of one of them, the curve of genus 2 may also be taken to be rational. For let the elliptic curves be

$$Y^2 = (X - a_1)(X - a_2)(X - a_3)$$
$$= F(X) \tag{14.3.1}$$

and

$$V^2 = (U - b_1)(U - b_2)(U - b_3)$$
$$= G(U). \tag{14.3.2}$$

There are rational r, s such that

$$(a_1 + r)(b_1 + s) = (a_2 + r)(b_2 + s) = (a_3 + r)(b_3 + s) = c \text{ (say)}, \tag{14.3.3}$$

since we have to solve only a pair of simultaneous linear equations. Clearly the curve

$$S^2 = F(T^2 - r) = \frac{-T^6}{\prod(b_j + s)} G\left(\frac{c}{T^2} - s\right) \tag{14.3.4}$$

does what is required.

It appears that in the more elaborate cases of reducibility referred to in Section 0, there are larger torsion subgroups of the two elliptic curves which are identified [Frey & Kani (1991), Prop 1.5].

4   Trying to show that a given jacobian is simple. There are now several examples in the literature of specific curves, defined over $\mathbb{Q}$, whose jacobians have been shown to be simple (= irreducible)¶ [Grant (1994c), Stoll (1995) and Flynn, Poonen & Schaefer (1995)].

¶ Whether or not the ground field k is algebraically closed, we shall always take the terms reducible and simple to be defined over the algebraic closure. When we say that the jacobian of a curve of genus 2 is simple, we shall always mean that there does not exist an isogeny over $\bar{k}$ to the product of two elliptic curves (some authors use the phrase 'absolutely simple' for this).

Suppose that the jacobian is isogenous over $\overline{\mathbb{Q}}$, and hence over some finite extension K of $\mathbb{Q}$, to the product of two elliptic curves $\mathcal{E}_1, \mathcal{E}_2$. Then the standard technique is to look at the local L factors at a place $\mathfrak{p}$ of K above a prime p, where $p \nmid 2 \operatorname{disc}(F)$, and to deduce consequences from the fact that

$$L_{\mathfrak{p}}(J/K, s) = L_{\mathfrak{p}}(\mathcal{E}_1/K, s) L_{\mathfrak{p}}(\mathcal{E}_2/K, s). \tag{14.4.1}$$

It would be a substantial diversion to discuss the mechanics of all this. We shall satisfy ourselves with quoting the consequence found in [Stoll (1995)], which the reader may find useful; namely that either

$$a_p^2 - 4(b_p - 2p) \text{ is a square in } \mathbb{Q} \tag{14.4.2}$$

where¶

$$a_p = p + 1 - \#\widetilde{C}(\mathbb{F}_p),$$
$$b_p = \frac{1}{2} \#\widetilde{C}(\mathbb{F}_{p^2}) + \frac{1}{2} \big(\#\widetilde{C}(\mathbb{F}_p) \big)^2 - (p+1) \#\widetilde{C}(\mathbb{F}_p) + p, \tag{14.4.3}$$

or

$$X^4 - \frac{b_p - 2p}{p} X^3 + \frac{a_p^2 - 2b_p + 2p}{p} X^2 - \frac{b_p^2 - 2p}{p} X + 1 = 0 \tag{14.4.4}$$

is satisfied by some nth root of unity.

One can therefore show that the jacobian is simple if some prime p can be found, $p \nmid 2 \operatorname{disc}(F)$, such that $a_p^2 - 4(b_p - 2p)$ is not a square in $\mathbb{Q}$ and equation (4) is not satisfied by any nth root of unity. Note that it is only necessary to check $n \in \{1, 2, 3, 4, 5, 6, 8, 10, 12\}$.

Our recommended technique for trying to decide whether the jacobian is reducible is as follows. First, the curve may be in of the forms described in Theorem 1.1, or [as for the curve (2.1)] easily mapped to such a curve by a fractional linear transformation in X, with the usual related transformation of Y.† In this case the jacobian is reducible. Otherwise, choose a prime p not

---

¶ The notation a_p, b_p is standard. Cognoscenti will recognize these as the a_p, b_p satisfying $L_p(J/\mathbb{Q}, s)^{-1} = 1 - a_p p^{-s} + b_p p^{-2s} - a_p p^{1-3s} + p^{2-4s}$. In computing a_p, b_p, take care to count the point at infinity the correct number of times, as described immediately after (8.2.7).

† If one uses property 3 in Theorem 1.1, then there are only finitely many fractional linear transformations to check. If $\theta_1, \ldots \theta_6$ are the roots of $F(X)$, then choose an ordered triple $[i, j, k]$, where the i, j, k are distinct and $1 \leqslant i, j, k \leqslant 6$. Let μ_{ijk} be the fractional linear transformation which takes $\theta_i, \theta_j, \theta_k$ to $\infty, 0, 1$. One then looks at $\mu_{ijk}(\theta_\ell), \mu_{ijk}(\theta_m), \mu_{ijk}(\theta_n)$ and checks whether any of these is the product of the other two. If this never happens for any of the $6 \cdot 5 \cdot 4$ possible choices of μ_{ijk} [it is an exercise for the reader to reduce the number of these which need to be checked], then the curve cannot be transformed into any forms given in Theorem 1.1.

dividing $2\operatorname{disc}(F)$ and hope both that the expression in (2) is not a square in $\mathbb{Q}$ and that (4) is not satisfied by any nth root of unity (and repeat the process with a new prime if the first choice of prime proves unlucky). We do not claim that this is an algorithm, but in our experience it has proved successful so far.

For example, consider the curve $\mathcal{C}_2 : Y^2 = (X^2+1)(X^2+2)(X^2+X+1)$, for which the rank of the jacobian was shown to be 1 in Example 11.4.1. The smallest prime not dividing $2\operatorname{disc}(F)$ is $p = 5$, when we find that $\#\widetilde{\mathcal{C}}(\mathbb{F}_5) = 6$ and $\#\widetilde{\mathcal{C}}(\mathbb{F}_{5^2}) = 22$. Then $a_5 = 0$, $b_5 = -2$ and so the expression in (2) is 48, which is not a square in $\mathbb{Q}$. Unfortunately, the equation in (4) becomes

$$X^4 + \frac{12}{5}X^3 + \frac{14}{5}X^2 + \frac{12}{5}X + 1 = 0, \qquad (14.4.5)$$

which is satisfied by $X = -1$ (twice!), and so the situation is not resolved. Trying $p = 7$, we find $\#\widetilde{\mathcal{C}}(\mathbb{F}_7) = 10$ and $\#\widetilde{\mathcal{C}}(\mathbb{F}_{7^2}) = 42$. Then $a_7 = -2$, $b_p = -2$ and so the expression in (2) is 68, which again is not a square in $\mathbb{Q}$. Fortunately, this time the polynomial in (4) becomes

$$X^4 + \frac{16}{7}X^3 + \frac{22}{7}X^2 + \frac{16}{7}X + 1 = 0, \qquad (14.4.6)$$

which is clearly not satisfied by any nth root of unity. Hence the jacobian is simple. It follows that the isogenous jacobian of the curve $\widehat{\mathcal{C}}_2$, given in (11.4.2), must also be simple.

Similarly, the curve $Y^2 = X(X-1)(X-2)(X-5)(X-6)$, of Example 11.3.1, is not resolved by $p = 7$ (the smallest prime not dividing $2\operatorname{disc}(F)$), but can be shown to have simple jacobian using $p = 11$. The first curve in Stoll (1995), which we have given in (11.5.1), is not resolved by $p = 11$ (the smallest prime not dividing $2\operatorname{disc}(F)$), but can be shown to have simple jacobian using $p = 13$.

Chapter 15

The endomorphism ring

0   Introduction. The classical geometric literature on this topic is vast and we shall not even contemplate tackling it here. We refer the reader to Hudson (1905), Chapter XVIII for a taste of it. For more modern treatments, see Brumer (1995a,b), de Jong & Noot (1991), and Mestre (1991a). Instead, we merely give a few examples of situations in which the jacobian of a curve of genus 2 has endomorphism ring larger than $\mathbb{Z}$.

1   A few examples. We begin with a few degenerate examples, where the jacobian is reducible and the endomorphism ring is larger than $\mathbb{Z}$. Let $\mathcal{C}$ be the curve

$$\mathcal{C}: \quad Y^2 = f_0 + f_1 X + f_2 X^2 + f_3 X^3 + f_2 X^4 + f_1 X^5 + f_0 X^6 \qquad (15.1.1)$$

so that the array of coefficients of the sextic is palindromic. Then the map $(x, y) \mapsto (1/x, y/x^3)$ takes $\mathcal{C}$ to $\mathcal{C}$, and naturally induces a new involution in the endomorphism ring of the jacobian. Note that the jacobian is reducible, since we can change coordinates to S, T, where $X = (S + 1)/(S - 1)$ and $Y = T/(S - 1)^3$. The resulting equation is $T^2 = (\text{cubic in } S^2)$ and so satisfies Criterion (i) of Theorem 14.1.1 for being isogenous to the product of two elliptic curves.

One can also obtain multiplication by i by requiring that only odd powers of X occur,

$$\mathcal{C}: \quad Y^2 = f_1 X + f_3 X^3 + X^5, \qquad (15.1.2)$$

and then looking at $(x, y) \mapsto (-x, iy)$. Again, the jacobian of any such curve is isogenous to the product of two elliptic curves. This can be seen by mapping $(x, y) \mapsto (x/\theta, y/\theta^{3/2})$, where θ is a root of $f_1 + f_3 X^2 + X^4$, and then applying Criterion (iii) of Theorem 14.1.3.

For other classes of curves with nontrivial endomorphism rings, we can use Richelot's isogeny to get multiplication by $\sqrt{-2}$ and $\sqrt{2}$. As in (9.2.1), let

$$\mathcal{C}: \quad Y^2 = F(X) = G_1(X)G_2(X)G_3(X), \qquad (15.1.3)$$

where

$$G_j(X) = g_{j2}X^2 + g_{j1}X + g_{j0} \tag{15.1.4}$$

and

$$\widehat{C}: \quad \Delta Y^2 = H(X) = L_1(X)L_2(X)L_3(X). \tag{15.1.5}$$

We introduce the notation

$$[G_i, G_j] = G_i'G_j - G_iG_j' \tag{15.1.6}$$

so that L_1, L_2, L_3 of (5) can be expressed as

$$L_1 = [G_2, G_3], \quad L_2 = [G_3, G_1], \quad L_3 = [G_1, G_2]. \tag{15.1.7}$$

We also define

$$\Delta(G_i, G_j, G_k) = \begin{vmatrix} G_{i0} & G_{i1} & G_{i2} \\ G_{j0} & G_{j1} & G_{j2} \\ G_{k0} & G_{k1} & G_{k2} \end{vmatrix}. \tag{15.1.8}$$

In particular, the Δ of (5) is just $\Delta(G_1, G_2, G_3)$.

We try to make L_1, L_2, L_3 scalar multiples of G_1, G_2, G_3. Note that the only possible non-degenerate way of doing this is for each L_i to be a multiple of G_i (since otherwise there would be an instance of $[G_i, G_j]$ being a scalar multiple of G_i, which can be shown to give rise only to degenerate solutions). Therefore, without loss of generality, we might as well impose

$$G_3 = L_3 = [G_1, G_2], \tag{15.1.9}$$

so that C is already of the form

$$C: \quad Y^2 = G_1G_2[G_1, G_2]. \tag{15.1.10}$$

It will be helpful to have the following technical lemma.

Technical Lemma 15.1.1. $[[G_i, G_j], [G_j, G_k]] = -2\Delta(G_i, G_j, G_k) \cdot G_j.$

If we now impose the further condition

$$L_2 = \kappa G_2 \tag{15.1.11}$$

for some κ, then it will follow from the technical lemma that:

$$L_1 = [G_2, G_3] = [\frac{1}{\kappa}[G_3, G_1], [G_1, G_2]]$$
$$= -\frac{2}{\kappa}\Delta(G_3, G_1, G_2) \cdot G_1 = -\frac{2}{\kappa}\Delta \cdot G_1. \tag{15.1.12}$$

Hence:
$$\widehat{C}: \quad Y^2 = -2G_1G_2G_3. \tag{15.1.13}$$

Now, $\widehat{J}$ is isomorphic to J via the isomorphism

$$\psi: \{(x,y),(u,v)\} \longmapsto \{(x,y/\sqrt{-2}),(u,v/\sqrt{-2})\}. \tag{15.1.14}$$

The map $\psi \circ \phi$ gives a nontrivial endomorphism, defined over $k(\sqrt{-2})$, of the jacobian J of C. This differs from the previous examples since the endomorphism on the jacobian is not induced by a birational map on the curve. The above is the natural analogue of the class of elliptic curves: $Y^2 = X(X^2 + aX + a^2/8)$.

It only remains to describe the condition on G_1, G_2 which forces them to satisfy (11), given (7),(9). In fact, a bit of algebra shows that, G_1, G_2 satisfy (7),(9),(11) if and only if $2g_{12}g_{20} - g_{11}g_{21} + 2g_{10}g_{22} = 0$. So, in summary, the class of curves is

$$C: \quad Y^2 = G_1G_2[G_1,G_2], \text{ where } 2g_{12}g_{20} - g_{11}g_{21} + 2g_{10}g_{22} = 0. \tag{15.1.15}$$

A variation can be made to the above construction, which would not be available for elliptic curves. Namely, rather than requiring each L_i to be a scalar multiple of G_i, we instead require it to be a scalar multiple of $\ddot{G}_i$, where

$$\ddot{G}_i = g_{i0}X^2 + g_{i1}X + g_{i2}. \tag{15.1.16}$$

Without loss of generality, start with the condition

$$G_3 = \ddot{L}_3, \tag{15.1.17}$$

so that
$$C: \quad Y^2 = G_1G_2\ddot{L}_3. \tag{15.1.18}$$

If we now impose the further condition

$$L_2 = \kappa\ddot{G}_2 \tag{15.1.19}$$

for some κ, then it will follow from the technical lemma that:

$$L_1 = [G_2,G_3] = [\frac{1}{\kappa}\ddot{L}_2,\ddot{L}_3] = \frac{1}{\kappa}[-[\ddot{G}_3,\ddot{G}_1],-[\ddot{G}_1,\ddot{G}_2]]$$
$$= -\frac{2}{\kappa}\Delta(\ddot{G}_3,\ddot{G}_1,\ddot{G}_2)\cdot\ddot{G}_1 = \frac{2}{\kappa}\Delta\cdot G_1. \tag{15.1.20}$$

Hence:
$$\widehat{C}: \quad Y^2 = 2\ddot{G}_1\ddot{G}_2\ddot{G}_3. \tag{15.1.21}$$

Now, $\hat{J}$ is isomorphic to J via

$$\varphi : \{(x,y),(u,v)\} \longmapsto \{(1/x, y/\sqrt{2}x^3), (1/u, v/\sqrt{2}u^3)\}, \qquad (15.1.22)$$

and $\varphi \circ \phi$ gives a nontrivial endomorphism of the jacobian, but which is now defined over $k(\sqrt{2})$.

It only remains to describe the condition on G_1, G_2 which forces them to satisfy condition (19), given (7),(17). In fact, a bit of algebra shows that G_1, G_2 satisfy (7),(17),(19) if and only if $2g_{12}g_{22} - g_{11}g_{21} + 2g_{10}g_{20} = 0$. So, in summary, the class of curves is

$$\mathcal{C}: \quad Y^2 = G_1 G_2 \ddot{L}_3, \text{ where } 2g_{12}g_{22} - g_{11}g_{21} + 2g_{10}g_{20} = 0. \qquad (15.1.23)$$

We have looked at a few instances of (15) and (23), and the jacobian is always reducible in the same manner as the 'awful warning' curve (14.2.1). We suspect that the jacobian is always reducible. It should be straightforward to express the two elliptic curves $\mathcal{E}_1, \mathcal{E}_2$, of the isogenous product, in closed form (in terms of the g_{ij}), but we have not done so. We leave this to a reader more eager than the one on p. 66.

It is worth noting that there is an intersection between the curves in (15) and (23), namely:

$$\mathcal{C}: \quad Y^2 = X(X^4 - a) \qquad (15.1.24)$$

The jacobian of such a curve has multiplication by $\sqrt{-2}$, $\sqrt{2}$ (and i of course); it is a special case of (2), and so the jacobian is without doubt reducible.

2   Complex and real multiplication.

2   Complex and real multiplication. An elliptic curve has full complex multiplication when the endomorphism ring is an imaginary quadratic extension of $\mathbb{Z}$. The analogous Criterion for an abelian variety of dimension 2 (such as the jacobian of a curve of genus 2) to have *complex multiplication* is that it is simple, and the endomorphism ring can be written in the form $\mathbb{Z}[\alpha][\beta]$, where $\mathbb{Z}[\alpha]$ is a real quadratic extension of $\mathbb{Z}$ and $\mathbb{Z}[\alpha][\beta]$ is an imaginary quadratic extension of $\mathbb{Z}[\alpha]$. The most natural example is the class of curves of the form:

$$\mathcal{C}: \quad Y^2 = X^5 + a, \quad a \neq 0. \qquad (15.2.1)$$

The jacobian of any such curve has multiplication by ρ, a primitive fifth root of unity, induced by $X \mapsto \rho X$. That is to say,

$$\{(x,y),(u,v)\} \longmapsto \{(\rho x, y),(\rho u, v)\}. \qquad (15.2.2)$$

Indeed the endomorphism ring is $\mathbb{Z}[\rho]$, which is an imaginary quadratic extension of $\mathbb{Z}[(1+\sqrt{5})/2]$. The special case when $a = 1/4$ has been studied in Grant (1995) in the context of quintic reciprocity.

Note that all curves of the form (1) are birationally equivalent over a finite extension of the ground field, using a map $(X, Y) \mapsto (c^2 X, c^5 Y)$. So, all curves (1) over $\overline{\mathbb{Q}}$ are simple if any one of them is. For example, $Y^2 = X^5 - 1$ is simple since, when $p = 11$, (14.4.2) becomes a non-square (in fact, 80) and (14.4.4) becomes

$$ X^4 + \frac{16}{11}X^3 + \frac{26}{11}X^2 + \frac{16}{11}X + 1 = 0, \tag{15.2.3} $$

which is not satisfied by any nth root of unity.

An abelian variety of dimension 2 has *real multiplication* when it is simple and its endomorphism ring includes $\mathbb{Z}[\alpha]$, a real quadratic extension of $\mathbb{Z}$ (this of course includes all cases with complex multiplication). These are of particular interest, due to the generalized Shimura-Taniyama conjecture. This asserts that any abelian variety A with real multiplication, defined over $\mathbb{Q}$, is isogenous to a factor of $J_0(N)$, for some N, where $J_0(N)$ is the jacobian of the modular curve $X_0(N)$ classifying isogenies of degree N. In equation (6.2) of Brumer (1995a), there is the three-parameter family of curves of genus 2

$$ \begin{aligned} \mathcal{C}_{bcd} : Y^2 &+ \left(x^3 + x + 1 + x(x^2 + x)\right)y \\ &= b + (1 + 3b)x + (1 - bd + 3b)x^2 + (b - 2bd - d)x^3 - bdx^4, \end{aligned} \tag{15.2.4} $$

for which the jacobian's endomorphism ring includes $\mathbb{Z}[\alpha]$, $\alpha^2 + \alpha - 1 = 0$. The justification for this is rather technical. It is due to the fact that the jacobians are those of $\mathcal{D}/\tau$, where $\mathcal{D}$ is a certain curve of genus 6, acted on by the dihedral group $\langle \sigma, \tau \mid \sigma^5 = \tau^2 = 1, \sigma\tau = \tau\sigma^{-1} \rangle$. Therefore $\sigma + \sigma^{-1}$ induces multiplication by $\sqrt{5}$ on the jacobian of $\mathcal{D}/\tau$. The details will appear in Brumer (1995b), as will other examples of real multiplication.

Chapter 16

The desingularized Kummer

0   Introduction. In one formulation of the classical theory of the jacobian of a curve C of genus 2, there occurs a nonsingular surface S intermediate between the jacobian $J(C)$ and the Kummer K. This surface is birationally equivalent canonically both to K and to K^*, and is a minimum desingularization of each of them. The configuration links the abelian variety self-duality of the jacobian with a projective space duality for K. We shall give an exposition of this in Chapter 17.

In our context, when the ground field is not algebraically closed, things cannot be quite like that. The Kummer is not projectively self-dual and we cannot expect the dual of the jacobian, when appropriately defined, to be an abelian variety: it is, at best, a principal homogeneous space. Nevertheless, there is a nonsingular surface S, which is canonically a minimum desingularization of both K and K^*. It arises naturally from a further study of the algorithms of Chapter 6, and is a natural half-way house in the duplication process.

In this chapter, we define and study S. At the end, we find an unramified natural covering, which we would expect to be the natural arena for a generalized abelian group duality. This will be a subject of later investigation.

The results described in this chapter were noticed only in the course of preparing these prolegomena. We have thus had to leave a number of loose ends.

1   The surface. Let¶ $F(X) = \sum f_j X^j \in k[X]$ be of precise degree 6 and without multiple factors. Let $\mathbf{p} = (p_0, ..., p_5)$, where the p_j are indeterminates, and put $P(X) = \sum_0^5 p_j X^j$. Chapter 6 suggests that we look at the projective locus S of the $\mathbf{p}$ for which $P(X)^2$ is congruent to a quadratic in X modulo $F(X)$. Clearly $P(X)^2$ is congruent to a polynomial of degree at

---

¶ As always, k is the ground field. Things defined over k may be referred to as *rational*.

most 5, say

$$f_6^5 P(X)^2 \equiv \sum_{j=0}^{5} C_j X^j \bmod F(X), \tag{16.1.1}$$

where the C_j are quadratic forms in $\mathbf{p}$ with coefficients in $\mathbb{Z}[f_0, \dots, f_6]$. Hence

$$\mathcal{S}: \qquad C_5 = C_4 = C_3 = 0, \tag{16.1.2}$$

is the intersection of the three quadric surfaces $C_j = 0$ $(j = 5, 4, 3)$.

We first find an equivalent definition of $\mathcal{S}$ working over the algebraic closure. Let θ_j $(j = 1, \dots, 6)$ be the roots of $F(X)$, say

$$F(X) = f_6 \prod (X - \theta_j). \tag{16.1.3}$$

Put

$$\begin{aligned}
P_j(X) &= \prod_{i \neq j} (X - \theta_i) \\
&= F(X) / \big(f_6(X - \theta_j)\big) \\
&= X^5 + \big(\theta_j + (f_5/f_6)\big) X^4 \\
&\quad + \big(\theta_j{}^2 + (f_5/f_6)\theta_j + (f_4/f_6)\big) X^3 + \cdots
\end{aligned} \tag{16.1.4}$$

and

$$\varpi_j = P_j(\theta_j) = F'(\theta_j)/f_6. \tag{16.1.5}$$

The P_j span the vector space of polynomials of degree at most 5: in particular,

$$P(X) = \sum_j \pi_j P_j(X), \tag{16.1.6}$$

where

$$\pi_j = P(\theta_j)/\varpi_j. \tag{16.1.7}$$

It is readily verified that

$$P_i(X)P_j(X) \equiv \left\{ \begin{matrix} 0 \\ \varpi_i P_i(X) \end{matrix} \right\} \bmod F(X) \qquad \text{for} \left\{ \begin{matrix} j \neq i \\ j = i \end{matrix} \right\}. \tag{16.1.8}$$

Hence

$$P(X)^2 \equiv \sum \varpi_j \pi_j^2 P_j \bmod F(X). \tag{16.1.9}$$

Putting everything together,

$$\mathcal{S} = \bigcap_j \mathcal{S}_j, \tag{16.1.10}$$

where S_j : $S_j = 0$ and¶

$$
\begin{aligned}
S_0 &= & C_5 &= f_6^4 \sum_j \varpi_j \pi_j^2 \\
S_1 &= & f_6 C_4 - f_5 C_5 &= f_6^5 \sum_j \theta_j \varpi_j \pi_j^2 \\
S_2 &= f_6^2 C_3 - f_6 f_5 C_4 + (f_5^2 - f_4 f_6) C_5 &= f_6^6 \sum_j \theta_j{}^2 \varpi_j \pi_j^2 .
\end{aligned}
\qquad (16.1.11)
$$

Here the S_j are quadratic forms in $\mathbf{p}$ with coefficients in $\mathbb{Z}[f_0, \ldots, f_6]$.

2   Elementary properties.

(i)   The π_j occur in (1.11) only as squares. Over the algebraic closure we may change the signs of the π_j: that is there is a unique $\widehat{P}(X)$ such that

$$
\widehat{P}(\theta_j) = \epsilon_j P(\theta_j) \qquad (16.2.1)
$$

for any choice of $\epsilon_j = \pm 1$. This gives $2^{6-1} - 1 = 31$ commuting involutions of the projective locus S.

(ii)   The $\mathbf{p} = (p_0, p_1, 0, 0, 0, 0)$ are clearly in S and form a rational line. Acting on it by the involutions gives 31 further lines, defined in general only over $\bar{k}$.

(iii)   The expressions (1.11) in terms of the coordinates π_j show that the S_j intersect transversally. Hence S is nonsingular.

(iv)   The projective equivalence class of the surface S depends only on the birational equivalence class of the curve C. This is clear for a transformation $X \mapsto X + \text{constant}$ in the model $Y^2 = F(X)$, so we need look only at $X \mapsto X^{-1}$ (with the usual associated transformation on Y, namely $Y \mapsto Y/X^3$). We may suppose that $F(0) \neq 0$, so that X is a unit modulo $F(X)$. Let $P(X) \in k[X]$ of degree 5 belong to a point of S, so $P(X)^2 \equiv H(X) \bmod F(X)$ where $H(X)$ is of degree 2. Define $Q(X) \in k[X]$ of degree 5 by $Q(X) \equiv X^4 P(X) \bmod F(X)$. Put $T = X^{-1}$. Then $P_T = T^5 Q(T^{-1})$, $H_T = T^2 H(T^{-1})$ and $F_T = T^6 F(T^{-1})$ are in $k[T]$ of respective degrees $5, 2, 6$. It is easily checked that $P_T^2 \equiv H_T \bmod F_T$ in $k[T]$. The transformation from the coefficients of $P(X)$ to those of P_T is clearly linear and invertible.

(v)   Finally, the quadric surfaces S_0 and S_2 are dual with respect to S_1:

¶ A similar, but not identical, form of the desingularization is given by Klein (1870) in his Inauguraldissertation. Cf. Chapter 17, Section 4.

that is, the dual¶ of a point of S_0 with respect to S_1 is tangent to S_2 and *vice versa*. Thus the configuration described in Section 1 is in a sense self-dual, though the duality does depend on the choice of model.

3   Equivalence with $\mathcal{K}$. Following Chapter 6 we now determine the map from $\mathcal{K}$ to S. It will be regular and birational, but not biregular.

Let $\mathfrak{X} = \{(x,y),(u,v)\}$ be a general element of Pic^2. There is a unique $M(X)$ of degree 3 such that $Y = M(X)$ meets C twice in $\mathfrak{X}$. Then

$$M(X)^2 - F(X) = (X - x)^2(X - u)^2 H(X) \qquad (16.3.1)$$

for a quadratic $H(X)$, where the divisor† $2\mathfrak{X}$ is given by $H(X) = 0$, $Y = -M(X)$. There is a unique polynomial $P^* = P^*(X)$ of degree at most 5 in X such that

$$(X - x)(X - u)P^*(X) \equiv M(X) \bmod F(X), \qquad (16.3.2)$$

so

$$P^*(X)^2 \equiv H(X) \bmod F(X). \qquad (16.3.3)$$

The polynomial $P^*(X)$ is not quite what we want. The coefficients of $P^*(X)$ are symmetric in (x,y) and (u,v) and so are in the function field of the jacobian. They are, however, odd under the involution of the function field induced by $Y \mapsto -Y$, as is easy to see and as we shall confirm shortly. If η is any odd element of the function field, for example $\eta = \alpha_0$, the coefficients of $P(X) = \eta P^*(X)$ are even, and so are in the function field of the Kummer. Clearly $P(X)$ satisfies the defining equations (1.2) of S. This is the required map from $\mathcal{K}$ to S.

It is plausible that the node $(0,0,0,1)$ of $\mathcal{K}$ blows up into the line on S of the $P(X)$ linear in X. The remaining 15 nodes blow up into its transformations under the automorphisms (2.1) with $\prod \epsilon_j = 1$.

¶ We are assured that this notion needs to be explained for the benefit of American graduate students. Let $S : S(\mathbf{X}) = 0$ be a regular (= nonsingular) quadric surface in $\mathbb{P}^n$, where $\mathbf{X} = (X_0, \ldots, X_n)$, and let $P(\mathbf{X}) = (\partial S/\partial X_0, \ldots, \partial S/\partial X_n)$. If $\mathbf{x}$ is on S, then $P(\mathbf{x})$ is the coordinates of the tangent at $\mathbf{x}$ to S. More generally, $\mathbf{X} \mapsto P(\mathbf{X})$ is a linear map from $\mathbb{P}^n$ to $\mathbb{P}^{n*}$ such that $\mathbf{x} \cdot P(\mathbf{y}) = \mathbf{y} \cdot P(\mathbf{x})$ (scalar products). This duality is known to generations of British schoolchildren under the name 'pole and polar'. See any sufficiently old-fashioned geometry text.

† We shall continue to use the identification between Pic^2 and Pic$^0 = J(\mathcal{C})$.

We can make the calculation explicit.¶ We have

$$
\begin{aligned}
M(X) = & \left(m_x(X-x)+1\right)\left((X-u)/(x-u)\right)^2 y \\
& + \left(m_u(X-u)+1\right)\left((X-x)/(u-x)\right)^2 v,
\end{aligned}
\tag{16.3.4}
$$

where m_x and m_u are determined by the conditions that the derivative of $F(X) - M(X)^2$ vanishes at $X = x$ and $X = u$. Hence

$$
m_x = \frac{F'x)}{2F(x)} - \frac{2}{x-u}.
\tag{16.3.5}
$$

We want to find an explicit $P^*(X)$ such that (2) holds. It is enough to multiply $M(X)$ by a scalar and to consider

$$
\begin{aligned}
M^\circ(X) = & \, 2(x-u)^3 yv M(X) \\
= & \left((F'(x)(x-u)-4F(x))(X-x)+2F(x)\right)(X-u)^2 v \\
& -\left((F'(u)(u-x)-4F(u))(X-u)+2F(u)\right)(X-x)^2 y.
\end{aligned}
\tag{16.3.6}
$$

Here all the terms on the right hand side are divisible by $(X-x)(X-u)$ except $F(x)(X-u)^2$ and $F(u)(X-x)^2$. But $F(X)-F(x) = (X-x)F(x,X)$ for some polynomial $F(x,X)$ and similarly for $F(X) - F(u)$. Hence

$$
M^\circ(X) \equiv (X-x)(X-u)P^\circ(X) \bmod F(X)
\tag{16.3.7}
$$

where

$$
\begin{aligned}
P^\circ(X) = & \left(F'(x)(x-u)-4F(x)-2F(x,X)\right)(X-u)v \\
& - \left(F'(u)(u-x)-4F(u)-2F(u,X)\right)(X-x)y.
\end{aligned}
\tag{16.3.8}
$$

This is of degree 6 in X, the coefficient of X^6 being $2f_6(y-v)$, so that $P^\triangle(X) = P^\circ(X) - 2(y-v)F(X)$ is of degree 5.

Putting it all together, the coefficients of $P^\triangledown(X) = P^\triangle(X)/(x-u)$ are odd elements of the function field of the jacobian, and $P^\triangledown(X)^2$ is congruent to a quadratic in X modulo $F(X)$.

¶ The rest of this section may be omitted at first reading.

4   Equivalence with $\mathcal{K}^*$. Again following Chapter 6, we determine the map from $\mathcal{K}^*$ to S. We use the machinery of Chapter 4, and so work over $\bar{k}$. The final results and formulae will be over k.

Let $\mathfrak{A}$ be an effective divisor of degree 3 in general position, say given by $U = 0$, $Y + W = 0$, where $U = U(X)$, $W = W(X) \in \bar{k}[X]$ are cubics in X. Hence

$$F = W^2 - UV, \tag{16.4.1}$$

where $V = V(X)$ is also a cubic.

By Riemann-Roch there are $S = S(X)$, $M = M(X) \in \bar{k}[X]$ of degree (at most) 1 and 4 respectively such that $SY + M$ meets $\mathfrak{A}$ with multiplicity 2. In particular, $-SW + M$ must be divisible by U, say $M = SW + IU$, where $I = I(X)$ is linear. Further, $S^2F - M^2$ must be divisible by U^2. On substituting for F and M this is equivalent to the condition that $SV + 2IW$ be divisible by U. We need

Lemma 16.4.1. *Let* $U, V, W \in \bar{k}[X]$ *be of degree* $\leqslant 3$. *Then there are* $R, S, T \in \bar{k}[x]$ *of degree* $\leqslant 1$ *and not all 0 such that* $RU + SV + TW = 0$. *Suppose, further, that* U, V, W *do not have a common zero and that one at least is of precise degree 3. Then* R, S, T *are unique up to a common multiplier in* $\bar{k}^*$.

The proof of the lemma is an easy exercise for the reader. In the case under discussion it is convenient to put $T = 2I$, so

$$RU + SV + 2IW = 0. \tag{16.4.2}$$

One checks that

$$S^2F - (SW + IU)^2 = GU^2, \tag{16.4.3}$$

where

$$G = RS - I^2 \tag{16.4.4}$$

is quadratic in X. It follows that the residual intersection $\mathfrak{D}$ is given by

$$\mathfrak{D}: \quad \begin{cases} G(X) = RS - I^2 = 0, \\ Y = -M/S = -(SW + IU)/S. \end{cases} \tag{16.4.5}$$

By the theory of Chapter 4, the residual intersection depends only on the linear equivalence class of $\mathfrak{A}$, as is easy to check. In the first place, on substituting $(V, U, -W)$ for (U, V, W), the first line of (5) is unchanged and the second is replaced by $Y = (RW + IV)/R$. By (2) this is the same on $G(X) = 0$. One now checks readily that the residual intersection is unchanged on replacing (U, V, W) by $(U, V + \lambda U, W)$ or by $(U + \lambda V, V, W)$

for any $\lambda \in \bar{k}$. In particular, the residual intersection $\mathfrak{D}$ is rational if the equivalence class of $\mathfrak{A}$ is rational.

If U, V are coprime, so $I \neq 0$, there is a formulation more symmetric in U and V. Let $\mathfrak{B}$ be the divisor $V = 0$, $Y - W = 0$, which is equivalent to $\mathfrak{A}$. Then $IY + J = 0$ passes through both $\mathfrak{A}$ and $\mathfrak{B}$ when

$$
\begin{aligned}
J &= IW + RU \\
&= -IW - SV \\
&= (RU - SV)/2
\end{aligned}
\tag{16.4.6}
$$

and $\mathfrak{D}$ is given by $G = 0$, $Y = -J/I$.

It remains to construct a map from Pic^3 to $\mathcal{S}$. By (3) we have

$$
GU^2 \equiv (SW + IU)^2 \bmod F.
\tag{16.4.7}
$$

Hence the $P(X)$ of degree 5 determined by

$$
UP \equiv -SW - IU \bmod F
\tag{16.4.8}
$$

satisfies

$$
P^2 \equiv G \bmod F.
\tag{16.4.9}
$$

By (2) and since $W^2 \equiv UV \bmod F$, an equivalent definition of P is

$$
VP \equiv RW + IV \bmod F.
\tag{16.4.10}
$$

It is easily seen that (8) and (10) are equivalent to

$$
WP \equiv J \bmod F,
\tag{16.4.11}
$$

where J is given by (6). Although the discussion leading to (6) required U and V to be coprime, this condition is not needed to prove the equivalence of (8), (10) and (11).

Presumably the 16 nodes of $\mathcal{K}^*$ blow up into the 16 lines on $\mathcal{S}$ which are not blowups of the nodes of $\mathcal{K}$.

5   Comparison of maps. For convenience we first enunciate the upshot of the previous sections.

THEOREM 16.5.1. *There is a map $\kappa : \mathcal{K} \to \mathcal{S}$ defined for general $\xi \in \mathcal{K}$ as follows:*
     Let $\mathfrak{X} = \{(x, y), (u, v)\}$ correspond to ξ. Put $G(X) = (X - x)(X - u)$ and let $M(X)$ be the cubic determined by the property that $Y - M(X)$ vanishes twice on $\mathfrak{X}$. Let $P(X) = \sum_0^5 p_j X^j$ be determined by $GP \equiv M \mod F$. Then $\kappa(\xi)$ is the point with projective coordinates $(p_0, \ldots, p_5)$.

THEOREM 16.5.2. *There is a map $\kappa^* : \mathcal{K}^* \to \mathcal{S}$ defined for general $\eta \in \mathcal{K}^*$ as follows:*
     Let $F = W^2 - UV$ correspond to η, where $U, V, W \in \bar{k}[X]$ are of degree 3. Let $I, R, S \in \bar{k}[X]$ be linear not all 0 such that $RU + SV + 2IW = 0$. Put $J = IW + RU$ as in (4.6) and let $P(X) = \sum_0^5 p_j X^j$ be determined by $WP \equiv J \mod F$. Then $\kappa^*(\eta)$ is the point with projective coordinates $(p_0, \ldots, p_5)$.

     These are related by

THEOREM 16.5.3. *Let $\xi \in \mathcal{K}$ and $\eta \in \mathcal{K}^*$ be dual, that is η gives the tangent to $\mathcal{K}$ at ξ. Then $\kappa(\xi) = \kappa^*(\eta)$.*

Proof. Let M, G be as in Theorem 1, so

$$F = M^2 - G^2 H \qquad (16.5.1)$$

for some quadratic H. Let $H = H_1 H_2$, where the H_j are linear. By Sextion 4 of Chapter 4, η corresponds to $F = W^2 - UV$, where $W = M$, $U = GH_1$, $V = GH_2$. We may thus take $I = 0$, $R = H_2$, $S = -H_1$, so $J = H_2 G H_1 = GH$. Let P_1, P_2 be the polynomial P in Theorem 1, 2 respectively. Then $GP_1 \equiv M \mod F$ and $MP_2 \equiv GH \mod F$, so (1) gives $P_1 = P_2$, as required.

6   Further possible developments. As already explained, this chapter describes work in progress.
     We should find descriptions of the maps κ, κ^* of Theorems 5.1, 5.2 as explicit expressions in the coordinates $(\xi_1, \ldots, \xi_4)$ and $(\eta_1, \ldots, \eta_4)$.
     We need to find an interpretation of the duality¶ mentioned at the end of Section 2.

¶ In the light of Chapter 17 it must be related to the situation discussed in Hudson (1905), Section 31.

The surface S has a double covering $\mathcal{J}$ (say) constructed as follows. The surface S comes from the behaviour of the six tropes T_j under the duplication map. More precisely, the ratios of the $T_j(2\mathfrak{X})$ are squares of functions of $\mathfrak{X}$ up to constants. It was shown in Chapter 6, Section 7 that this extends to all the 16 tropes. Let $\mathcal{J}$ be the surface in $\mathbb{P}^{15}$ related to all the tropes in the same way that S is related to the T_j. Clearly $\mathcal{J}$ is biregularly equivalent to the jacobian. It seems plausible that $\mathcal{J}$ is the appropriate arena for a generalization of the abelian variety duality of the jacobian and of the Poincaré divisor.

It is reasonable to hope that the 2-Selmer group can be investigated by 'twisting' $\mathcal{J}$ (or S) by cocycles.

Chapter 17

A neoclassical approach¶

0   Introduction. Kummer introduced his surface in an optical context and in terms of a quadric line complex, that is the set of lines in $\mathbb{P}^3$ satisfying a certain algebraic condition.† This is a powerful approach to many of the surface's properties, as Hudson (1905) shows. The connection with abelian varieties came later.

Fairly recently these two approaches have been unified in the classical context over the complexes by the description of the jacobian of a curve of genus 2 as the variety of lines on the intersection of two quadric surfaces in $\mathbb{P}^5$. This generalizes the description of a curve of genus 1 as the intersection of two quadric surfaces in $\mathbb{P}^3$ and is a special case of a more general result [Donagi (1980)]: it appears to have occurred first in Narasimhan & Ramanan (1969) and independently in Miles Reid's unpublished Cambridge thesis [Reid (1972)]. The description of the Kummer in terms of a quadric line complex comes by specializing one of the two quadric surfaces in $\mathbb{P}^5$.

In our context, we have to make a restriction on the curve $\mathcal{C}$ of genus 2, but it is fairly mild. We require the existence of a rational non-Weierstrass point. This we put 'at infinity'. We shall therefore be concerned with the curve

$$\mathcal{C}: \ Y^2 = F(X)$$
$$= f_6 X^6 + f_5 X^5 + \ldots + f_0, \tag{17.0.1}$$

with

$$f_6 = 1. \tag{17.0.2}$$

With this restriction, we shall show that by appropriate choice of the second quadric surface in $\mathbb{P}^5$ one obtains the jacobian and the Kummer in the form in which they appear elsewhere in this book. So far as the Kummer is concerned, the condition (2) is no real restriction because the Kummer is unchanged by the multiplication of $F(X)$ by a constant.

¶ This chapter is a reworking of the paper [Cassels (1993)]. We are grateful to the Cambridge Philosophical Society for permission to reproduce it. The notational conventions differ in places from those of the rest of the book.

† Definition later.

The account given here depends heavily on that in the last chapter of Griffiths & Harris (1978), though our aims are quite different. They want to display the power and potential of complex analysis: we hope to show that it is clearer to do without.

In the next section we prove

THEOREM 17.0.1. *Let Q_1, Q_2 be quadratic forms in six variables such that $Q_1 = 0$, $Q_2 = 0$ intersect transversely in $\mathbb{P}^5$. Let $\mathcal{W} \subset \mathbb{P}^5$ be the variety*

$$\mathcal{W}: \quad Q_1 = Q_2 = 0. \tag{17.0.3}$$

Denote by $\mathcal{A}$ the variety of lines on $\mathcal{W}$. Then $\mathcal{A}$ has a natural structure of principal homogeneous space over the jacobian of the curve $\mathcal{E}$ of genus 2 defined as follows. Let

$$E(\lambda_1, \lambda_2) = \det(\lambda_1 \mathbf{Q}_1 + \lambda_2 \mathbf{Q}_2), \tag{17.0.4}$$

where $\mathbf{Q}_j$ is the symmetric matrix for Q_j. Then, with an obvious convention, $\mathcal{E}$ is given by

$$\mathcal{E}: \quad E(\lambda_1, \lambda_2) = -\mu^2. \tag{17.0.5}$$

Note the $-$ sign in (5). That the two hypersurfaces intersect transversely is equivalent to saying that $E(\lambda_1, \lambda_2)$ has no multiple factors. If there is a rational ¶ line on $\mathcal{W}$, it can be taken as the neutral element of a group law on $\mathcal{A}$, which can then be identified with the jacobian of $\mathcal{E}$.

We now briefly sketch the rest of the argument.† First, take Q_2 to be the Grassmannian G of lines in $\mathbb{P}^3$. Then an element L of $\mathcal{A}$ corresponds to a pencil of lines in $\mathbb{P}^3$. The locus of the focus of the pencil is a surface $\mathcal{K}$, say. It is easy to see that the map from $\mathcal{A}$ to $\mathcal{K}$ is two-to-one, so $\mathcal{K}$ is the Kummer surface. The locus of the plane of the pencil is the projective-space dual of $\mathcal{K}$.

Note that $\det(G) = -1$, so the choice $Q_2 = G$ already implies that there is a rational non-Weierstrass point on $\mathcal{E}$

It remains to show that we can choose Q_1 so that $\mathcal{E}$ is any given curve $\mathcal{C}$ subject to $f_6 = 1$. We write a suitable Q_1 down; and the verification will be immediate. It will also be straightforward to check that the Kummer is in our standard form. In fact the Q_1 was obtained by 'reverse engineering' from the equation of the Kummer, and we explain briefly how it was done.

¶ By 'rational' we shall always mean 'defined over the ground field'.

† A reader who is unfamiliar with terms used here will find them defined later.

1　　Proof of theorem. We require some facts about quadric hypersurfaces, which we enunciate only for $\mathbb{P}^5$.

Lemma 17.1.1. *Let Q be a nonsingular quadratic form in six variables and let $\mathcal{Q}$ be the hypersurface $Q = 0$ in $\mathbb{P}^5$. The planes in $\mathcal{Q}$ fall into two algebraic classes. A line in $\mathcal{Q}$ lies in precisely one plane of each class. The classes of planes correspond naturally to the two solutions μ of*

$$\det(Q) = -\mu^2. \tag{17.1.1}$$

Proof. All this is standard,¶ so we shall sketch only a proof of the last sentence, following Hodge & Pedoe (1952), pp. 232–234. A plane Π in $\mathbb{P}^5$ is determined by the ratios of its Grassmann (= Plücker) coordinates. These are quantities p_S defined for all ordered triplets S of coordinate indexes and alternating in them. They are defined as follows: Let $\mathfrak{r}_j$ for $j = 1, 2, 3$ be a basis of the underlying affine space of Π. Then p_S is the minor belonging to S from the matrix whose columns are the $\mathfrak{r}_j$. The p_S are projective coordinates.

There is a dual set of line coordinates π_S defined in terms of the hyperplanes in which Π lies. Further,

$$\pi_S = \mu p_{S^*}. \tag{17.1.2}$$

for some constant μ, where S^* is such that SS^* is an even permutation of $\{1, \ldots, 6\}$.

If the plane Π spanned by the $\mathfrak{r}_j$ is in $\mathcal{Q}$, it is equally defined as the intersection of the three hyperplanes $\mathfrak{h}_j$, where

$$\mathfrak{h}_j = Q\mathfrak{r}_j \tag{17.1.3}$$

and Q is the matrix for Q. It follows from from (2) and (3) that there is a unique μ such that

$$\mu p_{S^*} = \sum_T p_T Q_{ST}, \tag{17.1.4}$$

where the sum is over triplets T and Q_{ST} is the 3×3 minor with rows from S and columns from T. Now (1) follows from (4) and some matrix algebra.

¶ For example, on $XU + YV + ZW = 0$ the line $Z = U = V = W = 0$ lies on the two planes $U = V = W = 0$ and $Z = U = V = 0$.

We shall also need to consider singular quadrics Q, but the situation is simpler:

Lemma 17.1.2. *Let Q be a quadratic form in six variables of rank 5, so Q is a cone: the notation otherwise being as in Lemma 1. Then the planes of Q form a single algebraic system. Each line of Q not through the vertex is on precisely one such plane.*

We now consider the pencil of quadratic forms

$$Q[\lambda_1, \lambda_2] = \lambda_1 Q_1 + \lambda_2 Q_2, \tag{17.1.5}$$

and the corresponding quadric hypersurfaces $\mathcal{Q}[\lambda_1, \lambda_2]$.

Lemma 17.1.3. *A plane Π can belong to at most one of the $\mathcal{Q}[\lambda_1, \lambda_2]$.*

Proof. For otherwise Π would be in all the quadric hypersurfaces. Take a basis for the underlying affine space of Π and extend it to a basis of the whole space. With this choice of basis, it is easy to see that (0.4) would be a perfect square, contrary to the hypothesis that it has no multiple factors.

We now define a pairing between the points

$$\mathfrak{a} = (\lambda_1, \lambda_2, \mu) \tag{17.1.6}$$

on $\mathcal{E}$ and the lines $L \in \mathcal{A}$ whose value is a line $\Upsilon(\mathfrak{a})L \in \mathcal{A}$. Suppose that $\mathfrak{a}$ and L are given. By Lemmas 1, 2 there is precisely one plane Π in $\mathcal{Q}[\lambda_1, \lambda_2]$ belonging to the family μ and containing L. But Π does not lie on all the quadrics of the pencil by Lemma 3, and so intersects $\mathcal{W}$ in L and another line M (say). Then $M = \Upsilon(\mathfrak{a})L$ by definition. Exceptionally, Π can be tangent to all the quadrics of the pencil, when $M = L$. Conversely, let $M \in \mathcal{A}$ meet L. Then by Lemma 3 there is precisely one quadric $\mathcal{Q}[\lambda_1, \lambda_2]$ of the pencil which contains the entire plane Π defined by L, M. Hence $M = \Upsilon(\mathfrak{a})L$, where $\mathfrak{a}$ is given by (6) and μ corresponds to the family of planes to which Π belongs.

We have thus proved

Lemma 17.1.4. *Let $\mathcal{A}_L$ be the set of lines $\Upsilon(\mathfrak{a})L$. Then $\mathcal{A}_L$ is the set of $M \in \mathcal{A}$ which meet L, with the above gloss on whether or not L belongs to it. Further, $\mathcal{A}_L$ is birationally equivalent to $\mathcal{E}$.*

Before defining the action of the jacobian of $\mathcal{E}$ on $\mathcal{A}$, it is convenient to have a further Lemma. We denote the automorphism $\mu \mapsto -\mu$ of $\mathcal{E}$ by a bar, $\mathfrak{a} \mapsto \bar{\mathfrak{a}}$, and will say that $\mathfrak{a}$, $\bar{\mathfrak{a}}$ are conjugate.

Lemma 17.1.5. Let $L_1, M_1, M_2 \in \mathcal{A}$ be distinct and suppose that L_1 meets M_1 and M_2. Then there is a unique further $L_2 \in \mathcal{A}$ which meets the M_j. Further, the points of $\mathcal{E}$ belonging to the planes Π_j containing L_j, M_j are conjugate.

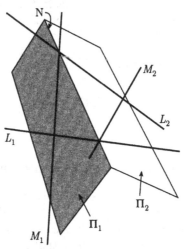

Proof. The existence and uniqueness of L_2 are easily proved by taking a basis of the coordinate system containing the intersections of L_1 with the M_j and further points on each of the M_j. Let $\mathcal{Q} = \mathcal{Q}[\lambda_1, \lambda_2]$ be the hypersurface of the pencil containing the plane Π_1 (say) spanned by L_1, M_1. Then, in particular, it contains the line N (say) joining the intersections of L_1, M_2 and of L_2, M_1. But $\mathcal{Q}$ contains the lines L_2, M_2 by the definition of $\mathcal{A}$, and so it contains the entire plane Π_2 spanned by them. The planes Π_j have the line N in common but are not identical: and so they belong to opposite systems on $\mathcal{Q}$. Hence they correspond to conjugate points of $\mathcal{E}$.

We are now ready to discuss the action of the jacobian of $\mathcal{E}$ on $\mathcal{A}$. As usual, we operate through the effective divisors of degree 2. All the divisors of the type $\{\mathfrak{a}, \bar{\mathfrak{a}}\}$ are linearly equivalent to each other: we denote their equivalence class by $\mathfrak{D}$. Otherwise, every divisor class $\mathfrak{A}$ of degree 2 contains precisely one effective divisor $\{\mathfrak{a}_1, \mathfrak{a}_2\}$.

We first give a provisional definition of the action, which works in the generic situation. Let $\mathfrak{A} = \{\mathfrak{a}_1, \mathfrak{a}_2\}$, and let $L \in \mathcal{A}$. Put

$$L_1 = L, \quad M_1 = \Upsilon(\mathfrak{a}_1)L, \quad M_2 = \Upsilon(\mathfrak{a}_2)L. \tag{17.1.7}$$

Then, in the notation of Lemma 5 the result of the action is L_2. By Lemma 5 we have $L_2 = \Upsilon(\bar{\mathfrak{a}}_1)M_2$, and so

$$L_2 = \Upsilon(\bar{\mathfrak{a}}_1)\Upsilon(\mathfrak{a}_2)L. \tag{17.1.8}$$

But the right hand side of (8) is always defined: we take it as the formal definition of the action of $\{\mathfrak{a}_1, \mathfrak{a}_2\}$. Note that

$$\Upsilon(\bar{\mathfrak{a}}_2)\Upsilon(\mathfrak{a}_1) = \Upsilon(\bar{\mathfrak{a}}_1)\Upsilon(\mathfrak{a}_2) \tag{17.1.9}$$

as operators. It is now a formality to check that the operation of effective divisors of degree 2 so defined is associative and commutative. Finally, to find the action of $\{\mathfrak{a}, \bar{\mathfrak{a}}\} \in \mathfrak{D}$ we take $\mathfrak{a}_1 = \mathfrak{a}$, $\mathfrak{a}_2 = \bar{\mathfrak{a}}$ in (8). The result is $\Upsilon(\mathfrak{a})^2$, which is clearly the identity by the definition of $\Upsilon(\mathfrak{a})$. On identifying $\mathfrak{A}$ with the divisor $\mathfrak{A} - \mathfrak{D}$ of degree 0, we have the required action of the jacobian.

2   The Kummer surface.

As heralded in the introduction, we take Q_2 to be the Grassmannian of lines in $\mathbb{P}^3$, whose properties we briefly recall. If $\mathfrak{u} = [u_1, u_2, u_3, u_4]$ and $\mathfrak{v} = [v_1, v_2, v_3, v_4]$ are two distinct points, then the coordinates of the line defined by them are

$$p_{ij} = u_i v_j - u_j v_i, \tag{17.2.1}$$

so $p_{ji} = -p_{ij}$, $p_{ii} = 0$. We treat them as the coordinates of the point

$$\mathbf{p} = [p_{43}, p_{24}, p_{41}, p_{21}, p_{31}, p_{32}] \tag{17.2.2}$$

in $\mathbb{P}^5$. The locus of $\mathbf{p}$ is $G(\mathbf{X}) = 0$, or $\mathcal{G}$, say, where G is the Grassmannian

$$G(\mathbf{X}) = 2X_1X_4 + 2X_2X_5 + 2X_3X_6. \tag{17.2.3}$$

There is a concrete interpretation of Lemma 1.1 in this context. The points of $\mathcal{G}$ represent lines in $\mathbb{P}^3$. The two families of planes in $\mathcal{G}$ represent respectively points and planes in $\mathbb{P}^3$. Finally, a line L in $\mathcal{G}$ represents a pencil of lines in $\mathbb{P}^3$. By Lemma 1.1, the line L lies in one plane of $\mathcal{G}$ of each system. These represent the *focus* $\mathfrak{f}(L) \in \mathbb{P}^3$ of the pencil (i.e. the point through which all the lines of the pencil pass) and the *plane* $\mathfrak{h}(L) \in \mathbb{P}^{3*}$ of the pencil (i.e. the plane in which they all lie). This description is effectively self-dual: we could have started with a line as the intersection of two planes.

In the notation of the previous section, put

$$Q_1 = H, \quad Q_2 = G, \tag{17.2.4}$$

where G is as above and H is any other form satisfying the conditions of Theorem 0.1. We denote the hypersurface $H = 0$ by $\mathcal{H}$. Then the points of $\mathcal{W} = \mathcal{G} \cap \mathcal{H}$ represent the lines of $\mathbb{P}^3$ whose coordinates (2) satisfy

$$H(\mathbf{p}) = 0: \tag{17.2.5}$$

what is classically known as a *quadric line complex*.

Recall from the previous section that $\mathcal{A}$ is the set of lines L in $\mathcal{W}$. Such a line represents a pencil of lines of the complex in $\mathbb{P}^3$. When is $u \in \mathbb{P}^3$ the focus of such a pencil? Clearly L should lie in the plane Π (say) of $\mathcal{G}$ represented by the point u. This can happen only if the restriction of H to Π is degenerate¶ or, in other words, if Π is tangent to $\mathcal{H}$. When this happens, Π contains a second line of $\mathcal{A}$, possibly† L 'taken twice'. The map from $\mathcal{A}$ to $\mathbb{P}^3$ given by $f(L)$ is thus two-to-one. The image $\mathcal{K}$ is the Kummer surface.

A line in $\mathbb{P}^3$ is called a *singular line* if it belongs to more than one pencil of the complex (or to one pencil, 'taken twice'): that is if the point $\mathbf{x}$ in $\mathbb{P}^5$ representing it is on more than one line of $\mathcal{A}$. If the plane $\Pi \subset \mathcal{G}$ belongs to the focus of one of the pencils, we have just seen that it lies in the tangent hyperplane Ω to $\mathcal{H}$ at $\mathbf{x}$.

Lemma 17.2.1. *A necessary and sufficient condition that $\mathbf{x} \in \mathcal{W}$ represent a singular line is that the tangent hyperplane to $\mathcal{H}$ at $\mathbf{x}$ be also tangent (somewhere) to $\mathcal{G}$.*

Proof. Let Ω be the tangent hyperplane to $\mathcal{H}$ at $\mathbf{x}$.

(i) Suppose, first, that $\mathbf{x}$ represents a singular line. Let the plane $\Pi \subset \mathcal{G}$ represent the focus of a pencil containing it. As noted above, $\Pi \subset \Omega$. But Ω has projective dimension 4 and G vanishes on Π, which has projective dimension 2. Hence the restriction of G to Ω is singular; that is, Ω is tangent to $\mathcal{G}$.

(ii) Conversely, suppose that Ω is tangent to $\mathcal{G}$, say at $\mathbf{y}$. The points $\mathbf{x}$ and $\mathbf{y}$ are distinct because $\mathcal{G}$ and $\mathcal{H}$ intersect transversally, by hypothesis. The line in $\mathbb{P}^5$ through $\mathbf{x}$, $\mathbf{y}$ is in $\mathcal{G}$, and so, by Lemma 1.1, is in precisely one plane of $\mathcal{G}$ of each class. Let Π be the one corresponding to a point in $\mathbb{P}^3$. Then $\Pi \subset \Omega$. Hence $\Pi \bigcap \mathcal{H}$ is a pair of lines through $\mathbf{x}$ (or a line taken twice) by the definition of Ω.

Corollary 1. $\mathbf{x}$ *determines Π uniquely.*

Corollary 2. *A necessary and sufficient condition that $\mathbf{x} \in \mathcal{W}$ belong to a singular line is that it lie on the quadric hypersurface $\mathcal{S}$ given (in an obvious notation) by the symmetric matrix*

$$\mathbf{S} = \mathbf{H}\mathbf{G}^{-1}\mathbf{H}. \tag{17.2.6}$$

Proof. The dual coordinates of the tangent at $\mathbf{x}$ are $\mathbf{m} = \mathbf{H}\mathbf{x}$, which is tangent to $\mathcal{G}$ when $\mathbf{m}^T\mathbf{G}^{-1}\mathbf{m} = 0$.

¶ H cannot vanish identically on Π by Lemma 1.3.

† This happens when u is a node. We do not follow this case in detail.

Note 1. We show in Section 4 that the surface $\Sigma = \mathcal{S} \cap \mathcal{W}$ is nonsingular. The map from Σ to $\mathcal{K}$ which takes a singular line into the corresponding point of $\mathcal{K}$ is a minimal desingularization of the latter. [So it is Σ here which compares with the $\mathcal{S}$ of Chapter 16.]

Note 2. The surface Σ depends only on the pencil of quadratic forms, and not on the choice of H in it. If H is replaced by $H + \lambda G$, then S is replaced by $S + 2\lambda H + \lambda^2 G$.

The dual construction. Instead of mapping L into the point $\mathfrak{f}(L)$, we map it into the plane $\mathfrak{h}(L)$ considered as a point of $\mathbb{P}^{3*}$. The argument above is unchanged if $\mathbb{P}^3$ is replaced by $\mathbb{P}^{3*}$. The map from L to $\mathfrak{h}(L)$ gives a two-to-one map from $\mathcal{A}$ to a variety $\mathcal{K}^* \subset \mathbb{P}^{3*}$. We shall show that $\mathcal{K}^*$ is the projective dual of $\mathcal{K}$.

First, we find when a point u of $\mathcal{K}$ and a point $\mathfrak{k}$ of $\mathcal{K}^*$ are incident.¶ By hypothesis, u is the focus of the pencil of lines belonging to some $L \in \mathcal{A}$; and, similarly, $\mathfrak{k}$ is the plane of the pencil of lines belonging to some $M \in \mathcal{A}$. Clearly, u, $\mathfrak{k}$ are incident precisely when the two pencils have a line in common; that is when L, M meet.

Now let $\mathbf{x} \in \mathcal{W}$ represent a singular line, and let Ω be the tangent hyperplane to $\mathcal{H}$ at $\mathbf{x}$. By Lemma 1, Ω contains a plane Π_1 belonging to a point $u \in \mathbb{P}^3$. Similarly, Ω contains a plane Π_2 belonging to some $\mathfrak{k} \in \mathbb{P}^{3*}$. By definition, u, $\mathfrak{k}$ are in $\mathcal{K}$, $\mathcal{K}^*$ respectively. We shall show that they are a dual pair. There are two lines L_1, L_2 (say) in Π_1 through $\mathbf{x}$, so $\mathfrak{f}(L_1) = \mathfrak{f}(L_2) = u$. Hence u is a double point of the intersection of $\mathcal{K}$ with $\mathfrak{k}$ (considered as a plane in $\mathbb{P}^3$): that is $\mathfrak{k}$ is the tangent plane to $\mathcal{K}$ at u, as required.

3   A special case.

Lemma 17.3.1. *Put*

$$
\begin{aligned}
H = &- 4X_1X_5 - 4X_2X_6 \\
&- X_3^2 + 2f_5X_3X_6 \\
&+ 4f_0X_4^2 + 4f_1X_4X_5 \\
&+ 4f_2X_5^2 + 4f_3X_5X_6 \\
&+ (4f_4 - f_5^2)X_6^2.
\end{aligned}
\tag{17.3.1}
$$

Then

$$
\det(\lambda_1 H + 2\lambda_2 G) = -4(f_0\lambda_1^6 + \ldots + f_5\lambda_1\lambda_2^5 + \lambda_2^6).
\tag{17.3.2}
$$

¶ That is, u is on $\mathfrak{k}$ considered as a plane in $\mathbb{P}^3$.

Further, every quadric of the pencil contains the line

$$X_3 = X_4 = X_5 = X_6 = 0. \qquad (17.3.3)$$

Proof. An easy direct verification.

The line N (say) given by (3) is a rational element of $\mathcal{A}$. Hence $\mathcal{A}$ has a structure of abelian variety (not just principal homogeneous space) with N as neutral element. In fact, the corresponding Kummer $\mathcal{K}$ is precisely the Kummer of the rest of this book with $f_6 = 1$: for brevity we shall refer to it as the 'old Kummer'. It is not logically necessary to know how (1) was constructed, and we defer a discussion until the next section. It is easy to confirm that (1) does give the old Kummer. Let $\mathfrak{z} \in \mathbb{P}^3$. In the notation (2.2) the condition that a point $\mathbf{X}$ on the Grassmannian represent a line through $\mathfrak{z}$ is that it lie on the plane Π given by

$$
\begin{aligned}
z_1 X_6 - z_2 X_5 + z_3 X_4 &= 0 \\
z_1 X_2 + z_2 X_3 - z_4 X_4 &= 0 \qquad (17.3.4) \\
z_1 X_1 - z_3 X_3 + z_4 X_5 &= 0.
\end{aligned}
$$

It is then straightforward to check that the intersection of $H = 0$ with Π is singular precisely when $\mathfrak{z}$ is on the old Kummer.

We proceed¶ to endow $\mathcal{A}$ with its structure as jacobian of $\mathcal{C}$, starting with $\mathcal{A}_N$. Consider a plane Π containing N and contained in a quadric $Q = H + 2sG = 0$, where s is a parameter. Then Π is spanned by N and a point $\mathbf{v}$ of the shape $(0, 0, v_3, v_4, v_5, v_6)$: so $Q(X_1, X_2, Tv_3, Tv_4, Tv_5, Tv_6)$ vanishes identically in X_1, X_2, T. Consideration of the cross-terms gives $X_5 = sX_4$ and $X_6 = sX_5$. Hence we can take $\mathbf{v} = (0, 0, t, 1, s, s^2)$ for some t with

$$-t^2 + (2f_5 t^2 + 4s^3)t + 4f_0 + 4f_1 s + 4f_2 s^2 + 4f_3 s^3 + (4f_4 - f_5^2)s^4 = 0. \quad (17.3.5)$$

We note that if

$$2y = t - f_5 s^2 - 2s^3, \qquad (17.3.6)$$

then (s, y) lies on $\mathcal{C}$. We have to find a line $L = L(s, t)$ other than N which lies in Π and in all the quadrics of the pencil. It is enough to look at the intersection of Π with $G = 0$, and it is readily seen that L is the line through $(-s, 1, 0, 0, 0, 0)$ and $(0, -ts, t, 1, s, s^2)$. Hence $\mathcal{A}_N$ consists of just such lines.

We can now construct the line of $\mathcal{A}$ corresponding to a general divisor $\{(s_1, t_1), (s_2, t_2)\}$ on (5). It is the line M (say) of $\mathcal{A}$ other than N meeting

---

¶ The rest of this section may be omitted at first reading. For clarity we treat only the generic case.

both $L(s_1, t_1)$ and $L(s_2, t_2)$. There is no difficulty in principle in doing this, but the resulting formulae are at first blush rather complicated. We know however that M lies on the plane (4), where $\mathfrak{z}$ gives the coordinates of the divisor on the (old) Kummer: that is

$$z_1 = 1, \quad z_2 = s_1 + s_2, \quad z_3 = s_1 s_2, \quad z_4 = \beta_0, \qquad (17.3.7)$$

where

$$
\begin{aligned}
(s_1 - s_2)^2 \beta_0 = &- 2y_1 y_2 + 2f_0 + f_1(s_1 + s_2) + 2f_2 s_1 s_2 + f_3(s_1 + s_2)s_1 s_2 \\
&+ 2f_4 s_1^2 s_2^2 + f_5(s_1 + s_2)s_1^2 s_2^2 + 2s_1^3 s_2^3 \qquad (17.3.8)
\end{aligned}
$$

and y_1, y_2 are given by (6). One checks that this actually is the case. It follows from (4) that X_1, X_2, X_6 are determined by X_3, X_4, X_5. We need therefore look only at the latter and now we find that the intersection of M with $L(s_j, t_j)$ has coordinates

$$(X_3, X_4, X_5) = (t_j, 1, s_j) \quad (j = 1, 2). \qquad (17.3.9)$$

Hence M is the line on the plane (4) determined by the two points (9).

On expressing the t_j in (9) in terms of y_j by (6) we deduce easily that M contains the point

$$
\begin{aligned}
(X_3, X_4, X_5) = \big(&y_2(f_5 s_1^2 + 2s_1^3) - y_1(f_5 s_2^2 + 2s_2^3), \\
&y_2 - y_1, \ y_2 s_1 - y_1 s_2\big).
\end{aligned}
\qquad (17.3.10)
$$

This point is unchanged by changing the signs of y_1, y_2 simultaneously, and so must represent the singular line at the point (7) of $\mathcal{K}$. It lies on the surface Σ introduced in the notes after Lemma 2.1, Corollary 2.

We have now achieved an integrated description of $\mathcal{A}$ and the old Kummer.

4   The construction. We indicate briefly how (3.1) was obtained, using ideas from the Inauguraldissertation of Klein (1870). Over the algebraic closure we can reduce G and H simultaneously to diagonal form, say

$$
\begin{aligned}
G &= \xi_1^2 + \ldots + \xi_6^2 \\
H &= \theta_1 \xi_1^2 + \ldots + \theta_6 \xi_6^2
\end{aligned}
\qquad (17.4.1)
$$

where the eigenvalues θ_j are distinct.

The quadratic form S of Lemma 2.1, Corollary 2 now takes the shape

$$S = \theta_1^2 \xi_1^2 + \ldots + \theta_6^2 \xi_6^2. \qquad (17.4.2)$$

Hence the surface Σ parametrizing the singular lines is the transverse intersection of three quadric hypersurfaces, and so nonsingular. It is thus a desingularization¶ of both $\mathcal{K}$ and $\mathcal{K}^*$.

Changes of the signs of the ξ_j leave the above configuration in $\mathbb{P}^5$ invariant, and so must leave invariant the configuration in $\mathbb{P}^3$ which it describes. However, an odd number of sign-changes interchanges the two families of planes on $\mathcal{G}$, and so interchanges points and planes in $\mathbb{P}^3$.

Now for the construction of the special H. Lemma 4.5.1 gives linear maps (over the algebraic closure) taking the old Kummer into its dual. The underlying matrix is skew and in terms of (2.2) suggests the eigenvector $\mathbf{v} = \mathbf{v}(\theta)$, where θ is one of the θ_j and

$$
\begin{aligned}
v_1 &= 2f_6\theta^5 + 2f_5\theta^4 + 2f_4\theta^3 + 2f_3\theta^2 + 2f_2\theta + f_1 \\
v_2 &= 2f_6\theta^4 + 2f_5\theta^3 + 2f_4\theta^2 + f_3\theta \\
v_3 &= 2f_6\theta^3 + f_5\theta^2 \\
v_4 &= 1 \\
v_5 &= \theta \\
v_6 &= \theta^2
\end{aligned}
\qquad (17.4.3)
$$

It is easily verified that

$$G(\mathbf{v}(\theta)) = 2F'(\theta) \qquad (17.4.4)$$

This suggests using the $G(\mathbf{X}, \mathbf{v}(\theta))$ as eigenforms, where $G(\ ,\ )$ is the associated bilinear form. One checks that

$$\sum_\theta \frac{G(\mathbf{X}, \mathbf{v}(\theta))^2}{F'(\theta)} = 2G(\mathbf{X}). \qquad (17.4.5)$$

The required new form is

$$\sum_\theta \frac{\theta G(\mathbf{X}, \mathbf{v}(\theta))^2}{F'(\theta)}; \qquad (17.4.6)$$

which gives the form H of the enunciation when $f_6 = 1$. In general it produces the formula with F/f_6 instead of F.

¶ cf. (16.1.11)

It is an uncovenanted mercy that the sums

$$\sum_\theta \frac{\theta^n}{F'(\theta)} \tag{17.4.7}$$

are easily computed by equating powers of X^{-1} in the identity

$$\frac{1}{F(X)} = \sum_\theta \frac{1}{F'(\theta)(X - \theta)} \tag{17.4.8}$$

(cf. the theory of the different).

Chapter 18

Zukunftsmusik

0   Introduction. This chapter discusses outstanding problems and possible future developments, some of which we have already alluded to.

1   Framework. A more appropriate theoretical framework than the jacobian might be the quotient Pic/$\mathfrak{O}$ of the Picard group by the canonical divisor. This has a natural $\mathbb{Z}/2\mathbb{Z}$ grading. Its quotient by the $\pm Y$ involution unifies the Kummer and its dual, as Chapters 3 and 6 suggest that they should be unified. The operators $\Upsilon(\mathfrak{a})$ of Chapter 17, Section 1 suggest that there is an underlying noncommutative structure.

2   Principal homogeneous spaces. In the genus 1 theory, curves of genus 1 are principal homogeneous spaces over the elliptic curves (= abelian varieties of dimension 1). They are naturally the 'twists' of the elliptic curves by cocycles. The principal homogeneous spaces of order 2 give a way to study $\mathfrak{G}/2\mathfrak{G}$ without leaving the ground field. For genus 2, the desingularized Kummer of Chapter 16 gives an opportunity to construct the principal homogeneous spaces of order 2, which are now surfaces, in a way closely analogous to the genus 1 construction. They might be equally useful for the study of $\mathfrak{G}/2\mathfrak{G}$.

3   Duality. In both the geometric and the highbrow arithmetic theory of curves, the abelian variety duality of the jacobian has a central rôle. As we have several times observed, this duality does not carry over in a naïve way to our setup. We suspect that further study of the desingularized Kummer $\mathcal{S}$ of Chapter 16 will give an appropriate generalization of the Poincaré divisor.

4   Canonical height. It would be nice to have a concrete formulation for our setup. One potential benefit would be a method for finding generators of the Mordell-Weil group $\mathfrak{G}$, given generators for $\mathfrak{G}/2\mathfrak{G}$, imitating the strategy described in [Cremona (1992), pp. 55–58]. This should be considerably more efficient that the method of Chapter 12, Section 3. This is work currently in progress, but we do not quite yet have a computationally viable method.

For curves of genus > 1, Mumford (1965) used the interaction of the canonical height and the group structure of the jacobian to show that rational points are very rare. [There is an exposition in Lang (1983), Chapter 5.] As a bare statement, this is superseded by Faltings' Theorem, but Mumford's approach is much more down-to-earth than Faltings'. With an explicit theory of heights, it might give explicit results in explicit situations. It is perhaps not too much to hope that one might obtain an extension of Chabauty's Theorem, say to Mordell-Weil rank 2.

5   Second descents. For genus 1, the existence of a skew-symmetric form on the Tate-Shafarevich group III often gives a cheap way to complete the determination of the Mordell-Weil group when there are suspected elements of order 2 in III. The skew-symmetric form exists for abelian varieties of any dimension, but the only treatment in print is stratospherically highbrow [Milne (1986a)]. It would be of interest to have explicit formulae for jacobians of genus 2.

6   Other genera. Much of the material in this book is peculiar to genus 2, but there are relations with other genera. For genus 3 with reducible jacobian cf. Faddeev (1960) and Cassels (1985). Recently Schaefer (1995, 1996a,b) has considered several problems for curves of genus > 2: Klassen & Schaefer (1996) have treated curves of genus 3 with an automorphism of order 3 but where the jacobian is not necessarily reducible.

7   A challenge of Serre's. As an alternative to attacking curves of higher genus frontally, one can use unramified covers to link curves of different genera. Unramified covers arise naturally, and indeed unnoticed by earlier arithmeticians, from considerations of factorization. An unramified cover of a curve of genus 1 is also of genus 1, but an unramified cover of a curve of genus > 1 is of yet higher genus.

Serre has raised the question whether the only rational points on

$$\mathcal{D}: \quad X^4 + Z^4 = 17Y^4 \qquad\qquad (18.7.1)$$

are the obvious ones with $Y = 1$ and $X, Z \in \mathbb{Z}$ [Serre (1988), top of p. 67]. This is a curve of genus 3 whose jacobian is isogenous to the product of three elliptic curves, but unfortunately the ranks of the curves are too high for Chabauty's Theorem to apply. They are also too high for the method of Demyanenko, which Serre was concerned with at the time.

As we shall see, there is a curve¶ $\mathcal{E}$ of genus 5 which is an unramified cover of $\mathcal{D}$ in more than one way. Every rational† point of $\mathcal{D}$ lifts to a rational point on $\mathcal{E}$ in at least one way. The curve $\mathcal{E}$ is an unramified cover of a curve $\mathcal{C}$ of genus 2. A determination of the rational points on $\mathcal{C}$ (finite ineffectively by Faltings) gives the rational points on $\mathcal{E}$, and so on $\mathcal{D}$. We had hoped that the Mordell-Weil rank of $\mathcal{C}$ would be 1, so Chabauty would apply. In fact, though we cannot find two independent divisors, the rank of $\mathcal{C}$ appears to be 2.

We used this $\mathcal{C}$ as a test-bed in our thoughts about genus 2 curves until, at a late stage, we realized that its jacobian is reducible, so $\mathcal{C}$ is completely anomalous. The determination of the Mordell-Weil group of $\mathcal{C}$ reduces to a study of a genus 1 curve over a rather repulsive number field. This we have not attempted yet. There may however be some methodological interest‡ in explaining the relation between $\mathcal{D}$ and $\mathcal{C}$.

The curve $\mathcal{D}$ is a ramified cover of the elliptic curve

$$X^4 + Z^4 = 17y^2, \tag{18.7.2}$$

which is birationally equivalent to

$$\rho(\rho + 1)(\rho - 1) = 34\sigma^2 \tag{18.7.3}$$

with

$$\rho = \frac{Z^2 + 4X^2 + 17y}{(X - 2Z)^2},$$
$$\sigma = \frac{Z^3 + 2X^3 + (Z + 8X)y}{(X - 2Z)^3}. \tag{18.7.4}$$

There are thus three unramified coverings of order 2 of (2) obtained by taking the square roots of $\rho + 1$, ρ and $\rho - 1$ respectively. Each lifts to an unramified covering of $\mathcal{D}$. We shall consider only the first of these.

Equation (1) for $\mathcal{D}$ can be written

$$\{17Y^2 + (5X^2 - 4XZ + 5Z^2)\}\{17Y^2 - (5X^2 - 4XZ + 5Z^2)\}$$
$$= -2(2X^2 - 5XZ + 2Z^2)^2. \tag{18.7.5}$$

¶ In fact there are three such curves, together giving a cover of $\mathcal{D}$ of high genus. But we have concentrated attention on one of them.

† In this section the ground field is $\mathbb{Q}$.

‡ This is the usual formula when a mathematician lavishes time and trouble on a dud idea and wants to capitalize on it somehow.

Let (X, Y, Z) be a rational point on $\mathcal{D}$. Without loss of generality X, Y, Z are coprime integers. Then

$$X \not\equiv 0, \qquad Z \not\equiv 0 \quad \mathrm{mod}\ 17 \tag{18.7.6}$$

and, without loss of generality,

$$Z \equiv 0, \quad X \not\equiv 0, \quad Y \not\equiv 0 \quad \mathrm{mod}\ 2. \tag{18.7.7}$$

By taking $-Z$ for Z if need be, we may suppose that

$$2X^2 - 5XZ + 2Z^2 = (2X - Z)(X - 2Z) \equiv 0 \quad \mathrm{mod}\ 17. \tag{18.7.8}$$

It is now easily checked that the greatest common divisor of the two main factors on the left hand side of (5) is 34. Clearly $5X^2 - 4XZ + 5Z^2$ is positive and congruent to 1 modulo 4. Hence

$$17Y^2 + (5X^2 - 4XZ + 5Z^2) = 34R^2$$
$$17Y^2 - (5X^2 - 4XZ + 5Z^2) = -68S^2 \tag{18.7.9}$$
$$2X^2 - 5XZ + 2Z^2 = 34RS$$

for some integers R, S; that is

$$(X + Z)^2 = 9(R^2 + 2S^2) - 28RS,$$
$$(X - Z)^2 = R^2 + 2S^2 + 12RS, \tag{18.7.10}$$
$$Y^2 = R^2 - 2S^2.$$

The three equations (10) define a curve $\mathcal{E}$ of genus 5, which covers the curve

$$\mathcal{C}: \quad T^2R^4 = (9R^2 - 28RS + 18S^2)(R^2 + 12RS + 2S^2)(R^2 - 2S^2) \tag{18.7.11}$$

of genus 2. If we could determine all the rational points on $\mathcal{C}$ then, working back, we would have all the rational points on Serre's curve $\mathcal{D}$. We suspect that the only rational points on $\mathcal{C}$ are those with $S = 0$. This, together with $34RS = (2X - Z)(X - 2Z)$ [the third equation in (9)], would answer Serre's challenge.

Clearly $\mathcal{C}$ has rank at least 1. We can show that the rank is at most 2, and we suspect that it is exactly 2, so Chabauty's Theorem does not apply. As already remarked (Chapter 14, Section 2), $\mathcal{C}$ has a reducible jacobian, so the determination of the Mordell-Weil rank is a problem about elliptic curves over number fields. As we fear the worst, we have left it for a rainy day.

Appendix I

MAPLE programs

This appendix contains the heavily annotated texts of two programs which illustrate the computations described in Chapter 3. They can be input directly into the MAPLE Algebra program. (There would be no point in typing in the annotations!) Alternatively, the programs can be downloaded by anonymous ftp, as explained in the next appendix.

```
#          Kummer.pro
#
#  Anything on a line after a # is comment (like this).
#
# The object of this MAPLE program is to confirm the computations in
# Chapter 3, 'The Kummer surface'. The essential formulae are
# saved in a file 'Kit.sav'. This may be used to compute
# with an explicitly given curve of genus 2. An example of
# this use is given in the program 'Appl.pro'.
#
#          Part 1
#
# This part just sets up some of the standard functions
# for later use. Here X is normally an indeterminate and we
# denote a (usually generic) divisor by (x,y),(u,v).
# The MAPLE system requires some modification of the usual
# notation. Note that multiplication must always be explicitly
# written as *. Exponentiation is denoted by ^, so X^3 is the
# cube of X.
# We use FX to denote F(X), which will
# always be a sextic, the coefficient of X^j being fj.
# We use Fxu to denote F0(x,u).   We often use s for
# x+u, p for x*u and b for beta0. When we need homogeneity
# we take t for 1 so sigma0:sigma1:sigma2:beta0 is t:s:p:b.
P := x*u:          # temporary notation
S := x+u:          # ditto
R := [2,S,2*P,S*P,2*P^2,S*P^2,2*P^3]:      # ditto
```

```
Fxu := sum('f.i*R[i+1]',i=0..6):      # F0(x,u)
FX := sum('f.i*X^i',i=0..6):          # F(X)
Fx := subs(X=x,FX):                    # F(x)
Fu := subs(X=u,FX):                    # F(u)
No := simplify((Fxu^2-4*Fx*Fu)/(x-u)^2):    # Norm of beta0
                                            # times (x-u)^2
   # The following routine transforms a homogeneous polynomial in
   # x,u into a polynomial in s=x+u, p=x*u. Note that we cannot
   # use S,P here (cf. above) since MAPLE would automatically
   # replace them by the above values in terms of x,u. We want to go the
   # other way.
Sym := proc(X)
local Z,Z1,Z2,x1,u1,t;
   Z := X;
   while Z <> subs(x = x1,u = u1,Z) do
     Z1 := subs(x = t,u = 0,Z);
     Z2 := Z-subs(t = x+u,Z1);
     Z2 := simplify(Z2/x/u);
     Z := subs(t = s,Z1)+p*Z2
   od:
   Z
end:
   # We now construct the equation of Kummer, that is the equation
   # for b=beta0 over the field of x,u (or s,p).
Norm := `Norm`:    # Clears reserved word Norm for use as a variable.
Norm := Sym(No):       # Norm of beta0 times s^2-4*p = (x-u)^2.
K := (x-u)^2*b^2-2*Fxu*b+No:   # Temporary notation

   # Now express in terms of s=x+u,p=xu
K := Sym(K);

   # We now express the Kummer in terms of the coordinates
   # xx[1]=sigma0,xx[2]=sigma1,xx[3]=sigma2,xx[4]=beta0
   # where we have homogenized.
K := simplify(xx[1]^4*subs(s=xx[2]/xx[1],p=xx[3]/xx[1],b=xx[4]/xx[1],K));

   #
   #           Part 2
   #
   # This part confirms the calculations in Section 2 of
   # Chapter 3, 'The Kummer surface'
   #
   # In this part we suppose that F(X)=G(X)H(X), where G is
   # of degree 2. We evaluate the linear map of coordinates
```

```
# sigma0,sigma1,sigma2,beta0 induced by adding the divisor
# A: G(X)=0, Y=0 to the divisor X := (x,y),(u,v)
# Note that  sigma0(A):sigma1(A):sigma2(A) = g2:-g1:g0.
# At a later stage we take g2=1.
GX := sum('g.i*X^i',i=0..2);    # G(X) of degree 2, coefficients gj
HX := sum('h.i*X^i',i=0..4);    # H(X) of degree 4.
# We are now concerned with the special form of FX, which we
# shall call FFX
FFX := collect(HX*GX,X);
#
# Now consider the intersection of the curve with a cubic
# Y=MX. We take it first to pass through GX=0, Y=0.
MX := GX*(m1*X+m0);
# We now want to choose this cubic so that Y=MX passes through
# the points (x,y) and (u,v). To remain in polynomials, we introduce
# a denominator n
Gx := subs(X=x,GX);
Gu := subs(X=u,GX);
n := Gx*Gu*(x-u);   # G(x)G(u)(x-u)
# Now solve equations for m0,m1.
solve(subs(X=x,MX)=n*y,subs(X=u,MX)=n*v,m0,m1);
assign(");   # gives m0,m1 these values
m0 := collect(m0,[y,v]);    # A MAPLE necessity before finding the
m1 := collect(m1,[y,v]);    # coefficients below. Here functions are linear
MX := collect(MX,[y,v]);   # in y,v.
My := coeff(MX,y,1);   # coefficient of first power of y in M(X).
Mv := coeff(MX,v,1);   # ditto for v.
# We now want the analogues of Fx,Fu,Fxu for the form FFX=GX*HX
FFx := subs(X=x,FFX);
FFu := subs(X=u,FFX);
# But for FFxu it is more difficult
FFxu := Fxu;
for i from 0 to 6 do
   FFxu := subs(f.i=coeff(FFX,X,i),FFxu)
od:
# We now square MX. We can now eliminate y,v by substituting for
# y^2,v^2 and yv. Put y^2=FFx,v^2=FFu, and b for
# beta0: b=(FFxu-2yv)/(x-u)^2. Then 2yv=FFxu-b*(x-u)^2.
# Call MX^2 with these substitutions MX2.
MX2 := My^2*FFx+Mv^2*FFu+My*Mv*(FFxu-b*(x-u)^2);
# To get the residual intersection, we equate this to n^2*FFX
W := `W`:    # Clears reserved letter W for use as a variable.
```

```
W := MX2-n^2*FFX:
   # we don't print W as it is very complicated.
   # We now take out the factors that we know are there.
   # Again we don't print out the fairly complicated intermediate results.
   # (Note that ending with a : means that the result is not
   # printed out, ending with a ; means that it is)
W := simplify(W/GX):
W := simplify(W/((X-x)*(X-u))):
W := simplify(W/(x-u)^2):
W := simplify(W/(Gx*Gu)):
   # Apply Sym. This expresses in terms of s=x+u, p=xu
W := Sym(W):
W := collect(W,X):   # Prepares for next stage.
   # Now look at the coefficients. We store them in table U.
for i from 0 to 2 do
   U[3-i] := collect(coeff(W,X,i),[b,p,s])
od;
   #
   # We now homogenize.
for i from 1 to 3  do
   V[i] := simplify(t*subs(b=b/t,p=p/t,s=s/t,U[i])):
   V[i] := collect(V[i],[t,s,p,b]):
od:
   # Now the V[i] are linear forms in t,s,p,b=sigma0,sigma1,sigma2,beta0
   # We take them as a vector of variables labelled 1,2,3,4
   # and we note that the values of t:s:p
   # for the transformed point are V[1]:-V[2]:V[3] and may be taken to
   # equal them.
   # change the sign of V[2]
V[2] := -V[2]:
   # We have to find the b=beta0 coordinate of the transformed point. For
   # this we use that the divisor A is of order 2, and so
   # addition of A is an involution in the projective space.
   # We need to invoke the MAPLE matrix routines.
with(linalg):   # allows us to use more algebra routines.
A := matrix(4,4);   # defines A as a 4 by 4 matrix, entries
                 # specified below.
   # This is to be the matrix of the transformation addition of A
   # Put in the elements we know already
for i from 1 to 3 do
   A[i,1] := coeff(V[i],t):
   A[i,2] := coeff(V[i],s):
   A[i,3] := coeff(V[i],p):
   A[i,4] := coeff(V[i],b):
od:
```

```
# Finally we get the coefficients of the new beta0 by using that
# the square of the transformation is the identity in projective
# space: that is, the square of the matrix A is a multiple of the
# unit matrix. We first put in indeterminates for the missing values
for i from 1 to 4 do
  A[4,i] := a.i
od:
  # We now square the matrix and solve for the ai by equating suitable
  # off-diagonal elements of the square B=A^2 to 0.
B := multiply(A,A):
solve(B[1,2]=0,a2):
assign("):
solve(B[1,3]=0,a3):
assign("):
solve(B[1,4]=0,a4):
assign("):
solve(B[2,1]=0,a1):
assign("):
  # With these values, we simplify the matrix A giving the transformation
  # and, for good measure, also the matrix B: which should be a multiple
  # of the identity.
for i from 1 to 4 do
for j from 1 to 4 do
  A[i,j] := simplify(A[i,j]);
  B[i,j] := simplify(B[i,j])
od:
od:
  # We now display the matrices that have resulted
  # We have to say op( ) as they are arrays.
op(A);
op(B);   # As it is not so easy to read the entries we confirm that
  # matrix B is diagonal.
simplify(B[2,2]-B[1,1]);   # Should be 0.
simplify(B[3,3]-B[1,1]);   # ditto
simplify(B[4,4]-B[1,1]);   # ditto

#
#      Part 3
#
# Here we aim to construct the 'bilinear' forms from the
# matrix giving the transformations by points of order 2.
# This gives flesh to the arguments sketched in Section 4
# of Chapter 3, 'The Kummer surface'.
# Recall that we look at the products of pairs of the
```

\# resultant linear forms and then identify functions of
\# the divisor of order 2 with corresponding functions of
\# a general divisor.
\#
\# The first thing is to transform the matrix A constructed
\# above, which gives addition by a divisor of order 2.
\# The matrix is given in terms of the coeffs of G and H,
\# where G is quadratic (defining the divisor of order 2)
\# and H is of degree 4. We first transform to the coefficients of G
\# and those of F=GH, eliminating those of H. We shall also put
\# g2=1 and g1=−s=−(x+u) and g0=p=xu. To eliminate the h's
\# it is convenient to define a procedure

```
Subh := proc(W)
local WW;
    WW := W;
    WW := subs(h0=f2+s*h1-p*h2,WW);
    WW := subs(h1=f3+s*h2-p*h3,WW);
    WW := subs(h2=f4+s*h3-p*h4,WW);
    WW := subs(h3=f5+s*h4,WW);
    WW := subs(h4=f6,WW);
    WW := simplify(WW);
    WW
end;
```

    \# The matrix program supplies the entries as a matrix A.
    \# We need to introduce the Kummer coordinates of the
    \# divisor G=0,Y=0 and so replace g2,g1,g0 by 1,−s,p.
    \# The resultant matrix is called B,
    \# the matrix in terms of the f's, p,s

```
B := map((W->subs(g2=1,g1=-s,g0=p,W)),A);
```

    \# And now we eliminate the h's using the procedure defined just above.

```
B := map(Subh,B);
```

    \# We define some familiar things.
    \# We now want the divisor G=0 to be a generic divisor
    \# of order 2. Since we no longer need the original
    \# uses of x, u we use them as a pair of distinct
    \# zeros of FX. We need to express that they are distinct.

```
Bxu := simplify((Fx-Fu)/(x-u)):
Bxu := collect(Bxu,u):
```

    \# x is a root of FX of degree 6. u is another root of FX, so
    \# satisfies an equation Bxu of degree 5 over k(x).
    \# Here k=Q(f0,...,f6) is the ground field. We are concerned
    \# with the field of symmetric functions of x,u: which is of degree 15.

```
# By using Bxu we can reduce a symmetric poly in x,u to degree at most
# 4 in u. By symmetry it is of degree at most 4 in x and so of
# the shape s^i*p^j, with i+j at most 4. This gives 15 elements,
# and so is a basis of k(s,p) over k.
#
# The following program takes a symmetric function of x,u
# and makes the reduction.
Simp := proc(W)
local WW;
   WW := W;
   WW := collect(WW,u);
   WW := rem(WW,Bxu,u);
   WW := collect(WW,x);
   WW := rem(WW,Fx,x);
   WW := simplify(WW);
   WW
end;
   # We now compute the beta0 of the divisor (x,0),(u,0)
Bet := simplify((Fxu-Fx-Fu)/(x-u)^2);
   #and put it in s,p terms
Bet0 := Sym(Bet);        # Used in TryOn below
   # We also do the same with s*beta0,p*beta0 and beta0^2
Bet1 := Sym(Simp((x+u)*Bet));    # Used in TryOn below
Bet2 := Sym(Simp(x*u*Bet));       #  ditto
Bet3 := Sym(Simp(Bet^2));         #  ditto

   # Note that 1,s,p,s^2,s*p,p^2, and the above 4 are linearly
   # independent over k. (By inspection of the values just obtained.
   # See also the routine TryOn below.)
   # Any polynomial in s,p is equal to a linear form in the 15 basis
   # elements s^i*p^j with i+j at most 4. We need a routine to do
   # this. It goes via x,u

SimpSP := proc(W)
local WW;
   WW := W;
   WW := subs(s=x+u,p=x*u,WW);
   WW := Simp(WW);
   WW := Sym(WW);
   WW
end;
```

```
# apply this to the matrix B of coefficients
DD := map(SimpSP,B):
    # We also give a program to convert a linear form in the
    # base elements into a quadratic form  in z1=1,z2=s,z3=p,z4=beta0.
    # As remarked above, these are linearly independent.  As they don't
    # span, it is a check on the correctness of the working that
    # (in any given case) we get a quadratic form in the z's.
    # Uses values of beta0,s*beta0,p*beta0,beta0^2 obtained above.
TryOn := proc(W)
local WW,W1,W2,W3;
    WW := W;
    WW := collect(WW,p);
    W1 := 1/3*coeff(WW,p,4)/f6^2; # coeff gives the coefficient of p^4
            # which occurs only in beta0^2=Bet3. So replace by z4^2.
    WW := simplify(WW–W1*Bet3+W1*z4^2);
    WW := collect(WW,p);
    W1 := coeff(WW,p,2); # look for coeff or p^2*s^2 which [apart
    W1 := collect(W1,s);  # from Bet3 (now gone)] occurs only in Bet2.
    W1 := –coeff(W1,s,2)/f6;
    WW := simplify(WW–W1*Bet2+W1*z3*z4); # replace Bet2 by z3*z4.
    WW := collect(WW,p);
    W1 := coeff(WW,p,1);    # Now Bet1 characterized by p*s^3.
    W1 := collect(W1,s);
    W1 := –1/2*coeff(W1,s,3)/f6;
    WW := simplify(WW–W1*Bet1+W1*z2*z4);
    WW := collect(WW,s);
    W1 := –coeff(WW,s,4)/f6;
    WW := simplify(WW–W1*Bet0+W1*z4); # Bet0 characterized by s^4.
        # Note: not z1*z4. What is left should be a quadratic in s,p.
        # We now homogenize the z's.
    WW := z1^2*subs(z2= z2/z1,z3 = z3/z1,z4 = z4/z1,WW);
    WW := subs(s=z2/z1,p=z3/z1,WW);  # Now substitute ratios for s,p.
    WW := collect(simplify(WW),[z1,z2,z3,z4]);   # simplify forms
    WW
end;
    # We first make linear
    # forms out of the matrix DD, which we got by transforming the
    # A given by the Matrix program. The coefficients are called y1,..,y4.
    # To satisfy MAPLE's peculiar rules for sum we must unset j
j := 'j';
```

```
# We now construct the xii(A+D) which occur on the right hand
# side of equation (4.2). In the present notation, this is Y[i]
# which we define next.
# Here Fraktur A
# is a generic divisor, whose Kummer coordinates will be
# (y1,y2,y3,y4) and Fraktur D is the divisor (x,0),(u,0)
# of order 2.
for i from 1 to 4 do
  Y[i] := sum('y.j*DD[i,j]',j=1..4)
od:

# Now for the biquadratic forms themselves. These are forms
# Bij quadratic in the yi and the coordinates zi of a second generic point
# Fraktur B, say. We have to show that the Bij are uniquely determined
# by the demand that (4.2) holds when Fraktur B is specialized
# to Fraktur D.
# We shall put the Bij in a square matrix BB
BB := array(1..4,1..4);
# In the computation we use that matrix BB is symmetric
# We now use the routines we constructed above to find the Bij
#
# The following is a massive computation, so allow time and don't
# become impatient.
for i from 1 to 4 do
for j from i to 4 do
  BB[i,j] := simplify(TryOn(SimpSP(Y[i]*Y[j])));
  BB[j,i] := BB[i,j];
od:
od:

# And now we check that they actually are symmetric in the
# y's and the z's, although they are treated so very differently.
# This also checks that there is nothing left over in the applications
# of procedure TryOn.
for i from 1 to 4 do
for j from 1 to 4 do
  Temp := BB[i,j]:
  TempA :=
  subs(y1=t1,y2=t2,y3=t3,y4=t4,z1=s1,z2=s2,z3=s3,z4=s4,Temp):
  TempA :=
  subs(t1=z1,t2=z2,t3=z3,t4=z4,s1=y1,s2=y2,s3=y3,s4=y4,TempA):
  C[i,j] := simplify(Temp–TempA);
od;
od;   # Now for the test. Should be all 0.
op(C);
```

```
#
#              Part 4
#
# This part checks the computations of Section 5 of Chapter 3.
#
# The object of this part is to obtain the formulae
# for duplication. We use the bilinear forms constructed
# in Part 3 and set the two divisors involved
# equal. We check that the elements BB[i,j] with i,j not 4
# all vanish, as they should. The Kummer coordinates
# of the double point are then (BB[1,4],BB[2,4],BB[3,4],BB[4,4]/2).
# Note that the last element is halved.
#
# For the doubling map, we have to equate the two sets of
# variables occurring in BB. For later use, it will be convenient
# to have them as a vector rather than a collection of variables.
# We make a temporary procedure to make the substitution.
Temp := proc(W)
local WW;
   WW := W;
   WW := subs(y1=xx[1],y2=xx[2],y3=xx[3],y4=xx[4],WW);
   WW := subs(z1=xx[1],z2=xx[2],z3=xx[3],z4=xx[4],WW);
   WW
end;
   # Make the substitution
DD := map(Temp,BB):
   # Now the doubling map is given by the elements of DD which involve
   # the index 4. All the others should vanish, at least as consequences
   # of the equation of the Kummer. Check latter statement first.
DD[1,1];
DD[1,2];
DD[1,3];
DD[2,3];
DD[3,3];
DD[2,2];
   # We want to show that DD[2,2] and DD[1,3] vanish as
   # a result of the definition of the Kummer.
simplify(DD[1,3]+K);
simplify(DD[2,2]-2*K);
   # Should be 0

   # We now define a vector giving the coordinates of the double
   # of a generic point
```

```
Doub := array(1..4);
Doub[1] := 2*DD[1,4]:
Doub[2] := 2*DD[2,4]:
Doub[3] := 2*DD[3,4]:
Doub[4] := DD[4,4]:
   # We make a procedure to double a vector on the Kummer
   # Since we are substituting a vector we have to use eval
   #
   # The peculiarities of MAPLE mean that we have to act on the
   # elements of the vector separately
Double := proc(W)
local i,Vec;
   for i from 1 to 4 do
      Vec[i] := eval(subs(xx=W,Doub[i]));
   od;
   Vec
end;
   #
   #         Part 5
   #
   # The object of this part is to set up the machinery for the
   # operation of the addition-subtraction bilinear form on a
   # given curve Y^2=F(X). Here we work with generic coefficients f.i.
   # It is envisaged that in working with this program one
   # equates the coefficients to those of the special form one
   # is interested in, but at first we leave them generic.
   #
   # For the applications it will be convenient to work with
   # vectors, so we make the appropriate transformation of the
   # biquadratic forms.
   # the coordinates of the initial points
   # will be yy[1]...yy[4] and zz[1]...zz[4]
yy := array(1..4);
zz := array(1..4);
Temp := (WW ->subs(y1=yy[1],y2=yy[2],y3=yy[3],y4=yy[4],
z1=zz[1],z2=zz[2],z3=zz[3],z4=zz[4],WW)):
BBB := map(Temp,BB):
   # This is the final shape of the matrix of biquadratic forms
```

```
# The following routine computes the Kummer coordinates of a
# divisor (xx,yy),(uu,vv) as (X[1],X[2],X[3],X[4]).
Coord := proc(xx,yy,uu,vv)
local X,Temp;
   X[1] := 1;
   X[2] := xx+uu;
   X[3] := xx*uu;
   Temp := subs(x = xx,u = uu,Fxu);
   X[4] := simplify((Temp-2*yy*vv)/(xx-uu)^2);
   X
end:
```

```
# we now want to evaluate the Kummer of a vector. The procedure
# Kum does this.
Kum := proc(ww)
local Temp;
   Temp := subs(xx=ww,K);
   Temp := eval(Temp):
   Temp := simplify(Temp);
   Temp
end;
```

```
# We also need a procedure to evaluate the bilinear form for two
# given vectors.
Bilin := proc(vv,ww)
local Temp,Res;
   for i from 1 to 4 do
   for j from 1 to 4 do
     Temp := BBB[i,j];
     Temp := subs(yy=vv,zz=ww,Temp);
     Temp := eval(Temp);
     Temp := simplify(Temp);
     Res[i,j] := Temp;
   od;
   od;
   Res
end;
```

```
# To sum up. Kum( ) gives the Kummer and Bilin( , ) gives the
# bilinear form. To work with a given form one has to equate
# its coefficients with those (fi) of the generic form. The function
# Coord(x,y,u,v) computes the xi coordinates of the divisor {(x,y),(u,v)}
# on the curve. Now, save what we need in the file Kit.sav
save(FX,Fx,Fu,Fxu,K,BBB,Doub,Double,Coord,Kum,Bilin,`Kit.sav`):
```

```
#
#          Appl.pro
#
#          An application
#
# This program makes use of the formulae and procedures constructed
# in the programs Kummer to check the results of Section 6
# of the Chapter 3, 'The Kummer surface'.
# We consider a special curve with coefficients
f6 := 1; f5 := -1; f4 := 3; f3 := -5; f2 := 0; f1 := -10; f0 := 16;
# Note that the corresponding F(X) takes the square values
# 36,16,4,36 at -1,0,1,2 respectively.
#
# We now bring back the formulae we need.
# MAPLE automatically makes the above substitutions for the fj
# in the formulae
interface(quiet=true):    # cut the echo to screen
read `Kit.sav`:    # produced already by the program Kummer.pro
interface(quiet=false):    # restores the normal echo
# We first compute the xi coordinates of some divisors on the curve.
# First of the divisor (-1,6),(0,-4)
TT := Coord(-1,6,0,-4);
op(TT);
# Again consider the divisor (1,2),(2,-6)
SS := Coord(1,2,2,-6):
op(SS);
# Check that they do indeed lie on the Kummer
Kum(TT);
Kum(SS);
# Both should be 0

# Next check the doubling program
DDD := Double(TT);
op(DDD);
# There is a common factor, but we do not bother to take it out.
# check that we are on the Kummer
Kum(DDD);
# should be 0

# Now compute the corresponding bilinear form
Ory := Bilin(TT,SS):
op(Ory);
# By construction the elements of Ory should be (ai*bj+aj*bi)/2,
# where the ai and the bi are the Kummer coordinates of the
```

```
# sum and difference (but we don't know which is which). It is
# easy to devise a general procedure for finding the ai and the bi.
# In this case we get
PP := [0,11,17,99];
QQ := [24,31,-7,303];
# We check that they are actually on the Kummer
Kum(PP);
Kum(QQ);
# Here the support of PP is a rational point at infinity and
# a rational point with X=17/11. Check that there is such a rational
# point by substituting in FX
subs(X=17/11,FX);
ifactor(");
# so it is a rational square

# But the support of QQ is a pair of conjugate points over
# a quadratic field. Check. Denote the square root of 1633 by Om.
Om := RootOf(s^2=1633,s);
Psi := `Psi`:     # Clears reserved word Psi for use as a variable.
Psi := (31+Om)/48;  # Check that this is a root of the quadratic equation
simplify(24*Psi^2-31*Psi-7);
# So one of the conjugate points has X=Psi. Substitute in FX. We should
# get a square
Lam := simplify(subs(X=Psi,FX));
# factor the denominator
ifactor(382205952);
# So twice the denominator is a square
# multiply by this square
Lama := 2*382205952*Lam;
# Take out the square 25
Lamb := Lama/25;
# If this is a square, it is a multiple of (T-sqrt1633)^2 for some T
solve(71377625*(T^2+1633)-2*3996438025*T=0);
# Look at the two values of T in turn
simplify((28897-305*Om)^2);
# Only a non-square rational multiple of Lamb, so no good for us.
# So look at the other value of T
simplify((7015-407*Om)^2);
# Is equal to Lamb
```

Appendix II

Files available by anonymous ftp.

We have made available a number of files, some of which have been mentioned in the text. These can all be found at ftp.liv.ac.uk in the directory ~ftp/pub/genus2 by anonymous ftp. You should first type:

ftp ftp.liv.ac.uk

and then login using *anonymous* as your name, and your e-mail address as your password. Then type:

cd ~ftp/pub/genus2

which will take you to the correct directory. Once you have further changed to the desired subdirectory, then the standard command:

get *filename*

will copy the file *filename* to your local directory.

The current directory structure is as follows. In addition to the following files, there is also a file README in each directory, describing the contents of that directory. These should be read to check for any changes recently made to the directory. Please send any questions about these files to: evflynn@liv.ac.uk preferably with the subject header *genus 2 ftp files*.

Directory structure.

The main directory ~ftp/pub/genus2 has the subdirectories:

jacobian.variety , **kummer** , **local** , **maple,**
misc , **out.of.date** , **rank.tables**

Directory jacobian.variety

This directory contains files relating to the jacobian variety of a general curve of genus 2, $Y^2 = f_0 + \ldots + f_6 X^6$, using the embedding into $\mathbb{P}^{15}$ given by the coordinates $z_0, \ldots, z_{15}$ described at the beginning of Chapter 2, Section 2. The file **bilinear.forms** gives the 16 bilinear forms $\Phi_{ij} = \Phi_{ij}\big(\mathbf{z}(\mathfrak{A}), \mathbf{z}(\mathfrak{B})\big)$ described in Lemma 3.9.1. Recall that these are defined over $\mathbb{Z}[f_0, \ldots, f_6]$ and satisfy

$$\Big(\xi_i(\mathfrak{A} + \mathfrak{B})\xi_j(\mathfrak{A} - \mathfrak{B})\Big) = \Big(\Phi_{ij}\big(\mathbf{z}(\mathfrak{A}), \mathbf{z}(\mathfrak{B})\big)\Big).$$

The other file in the directory is **defining.equations**, which gives a set of defining equations for the jacobian variety, expressed as 72 quadratic forms. These give a basis for all quadratic relations.

Directory kummer

This directory contains files relating to the kummer surface, as embedded into $\mathbb{P}^3$ using the coordinates $\xi_1, \ldots, \xi_4$ described in (3.1.3). The file **defining.equation** gives the quartic equation which defines the Kummer surface; we have also included in this file the linear map given by addition by a point of order 2. The file **biquadratic.forms** gives the biquadratic forms B_{ij} described in Theorem 3.4.1. Recall that these are biquadratic in the $\xi_j(\mathfrak{A})$, $\xi_j(\mathfrak{B})$, defined over $\mathbb{Z}[f_0, \ldots, f_6]$ and satisfy

$$\big(\xi_i(\mathfrak{A} + \mathfrak{B})\xi_j(\mathfrak{A} - \mathfrak{B}) + \xi_i(\mathfrak{A} - \mathfrak{B})\xi_j(\mathfrak{A} + \mathfrak{B})\big) = 2\big(B_{ij}(\mathfrak{A}, \mathfrak{B})\big).$$

Directory local

This contains files relating to local power series in the kernel of reduction, as described in Chapter 7. The file **formal.group** gives the terms of the formal group $\mathcal{F}(\mathbf{s}, \mathbf{t})$ of (7.1.14) up to terms of degree 7 in the local parameters. Terms are also given in **local.coordinates** for the power series expansions $\sigma_i(\mathbf{s}) \in \mathbb{Z}_f[[\mathbf{s}]]$ of the local coordinates, described in (7.1.11). We also give up to degree 7 the expansions for the formal exponential and logarithm maps of Definition 7.2.2, in the files **exp** and **log** respectively. Note that the files in this directory, together with Φ_{ij} above, give everything necessary to apply Chabauty's Theorem, as described in Chapter 13.

Directory maple

This contains files of MAPLE programs. This includes **Kummer.pro**, **Appl.pro** which give all of the MAPLE programs in Appendix I. There is a file **group.law** which adds divisors on the jacobian. There is also the file **kernel.of.mu**, which constructs the polynomial $h(X)$ described in (6.5.4), making it easy to decide whether or not the map μ of (6.1.9) has kernel precisely $2\mathfrak{G}$. Further programs will at some stage be added to this directory, including implementations of the 2-descent techniques described in Chapter 11.

Directory misc

This contains files which don't seem to fit any other obvious category. The only file currently here is **people**, which has a list of people (including e-mail addresses) who are currently doing research into curves of genus 2. Anyone who would like to be included on the list should e-mail me at evflynn@liv.ac.uk and I'll add your name.

Directory out.of.date

Hardly anyone should have any reason to use this directory. There used to be an alternative, but now out-of-date, basis in $\mathbb{P}^{15}$, which was almost the same as $z_0, \ldots, z_{15}$ above, except for a trivial linear adjustment. This basis was used in Flynn (1990a,1993a,1994). To avoid the confusion of having two competing bases persisting, we have selected the one given in this book to be the standard from now on. This directory is merely the scrap heap for files which related to this other basis.

Directory **rank.tables**

This contains files of curves for which the rank of the jacobian has been found. At the moment these are all in the single file **ranks1**. However, as the number of known ranks increases, there will be a reorganization into several files.

Bibliography

Adelman, L.M. & Huang, M.A. (1992) *Recognizing primes in r andom polynomial time.* Springer Lecture Notes in Mathematics, **1512**.

Baker, H.F. (1907) *An introduction to the theory of multiply periodic functions.* Cambridge University Press.

Bavencoffe, E. *See* Boxall, J.

Berry, T.G. (1984) Detecting torsion divisors on curves of genus 2. *EUROSAM 84*, 108–114. *Springer Lecture Notes in Computer Science*, **174**.

Berry, T.G. (1992) Points at rational distance from the vertices of a triangle. *Acta Arith.* **62**, 391–398.

Bertin, J. (1988) Reseaux de Kummer et surfaces K3. *Invent. Math.* **93**, 267–278.

Bolza, O. (1888) On binary sextics with linear transformations into themselves. *Amer. J. Math.* **10**, 47–70.

Bombieri, E. (1990) The Mordell conjecture revisited. *Ann. Scuola Norm. Sup. di Pisa* (IV) **17**, 615–640. Corrections **18** (1991), 473.

Borchardt, C.W. (1877) Ueber die Darstellung der *Kummer*schen Fläche vierter Ordnung mit sechszehn Knotenpunkten durch die *Göpel*sche biquadratische Relation zwischen vier Thetafunctionen mit zwei Variabeln. *J. reine angew. Math.* **83**, 234–244.

Bost, J.-B. & Mestre, J.-F. (1988) Moyenne arithmético-géométrique et périodes des courbes de genre 1 et 2. *Gaz. Math. Soc. France* **38**, 36–64.

Bost, J.-B., Mestre, J.-F. & Moret-Bailly, L. (1990) Sur le calcul explicite des 'classes de Chern' des surfaces arithmétiques de genre 2. *Astérisque* **183**, 69–105.

Boxall, J. (1992) Valeurs spéciales de fonctions abéliennes. *Groupe d'étude sur les problémes diophantiens 1991-2. Publications mathématiques de l'université Pierre et Marie Curie*, **103**.

Boxall, J. & Bavencoffe, E. (1992) Quelques propriétés arithmétiques des points de 3-division de la jacobienne de $y^2 = x^5 - 1$. *Séminaire de théorie des nombres, Bordeaux* **4**, 113–128.

Brumer, A. & Kramer, K. (1994) The conductor of an abelian variety. *Compositio Math.* **92**, 227–248.

Brumer, K. (1995a) The rank of $J_0(N)$. *Astérisque* **228**, 41–68.

Brumer, K. (1995b) Curves with real multiplication. In preparation.

Cantor, D.G. (1987) Computing on the jacobian of a hyperelliptic curve. *Math. Comp.* **48**, 95–101.

Cartier, P. (1960) Isogenies and duality of abelian varieties. *Ann. Math.* **71**, 315–351.

Cassels, J.W.S. (1963) Arithmetic on curves of genus 1. (V) Two counter-examples. *J. London Math. Soc.* **38**, 244–248.

Cassels, J.W.S. (1983) The Mordell-Weil group and curves of genus 2. pp. 29–60 of: *Arithmetic and geometry. Papers dedicated to I.R. Shafarevich on the occasion of his sixtieth birthday.* Vol. I, *Arithmetic.* Birkhäuser, Boston, Mass.

Cassels, J.W.S. (1985) The arithmetic of certain quartic curves. *Proc. Roy. Soc. Edinburgh.* **100A**, 201–218.

Cassels, J.W.S. (1986) *Local fields.* LMS Student Texts **3**. Cambridge University Press.

Cassels, J.W.S. (1989) Arithmetic of curves of genus 2. pp. 27–35 of *Number theory and applications.* (ed. R.A. Mollin). NATO ASI Series **C,265**. Kluwer Academic Publishers, Dordrecht.

Cassels, J.W.S. (1991) *Lectures on elliptic curves.* LMS Student Texts **24**. Cambridge University Press.

Cassels, J.W.S. (1993) Jacobian in genus 2. *Math. Proc. Cambridge Philos. Soc.* **114**, 1–8.

Cassels, J.W.S. (1995) Computer-aided serendipity. *Rendiconti del Seminario Matematico dell'Università di Padova* **93**, 187–197.

Cayley, A. (1846) Sur la surface des ondes. *J. Math. Pures Appl.* (Liouville) **11**, 291–295. (= *Collected Mathematical Papers* **I**, 302–305.)

Cayley, A. (1877) On the double Θ-functions in connexion with a 16-nodal quartic surface. *J. reine angew. Math.* **83**, 210–219. (= *Collected Mathematical Papers* **X**, 157–165.)

Chabauty, C. (1941a) Sur les points rationnels des courbes algébriques de genre supérieur à l'unité. *C. R. Acad. Sci. Paris* **212**, 882–885. [MR 3–14].

Chabauty, C. (1941b) Sur les points rationnels des variétés algébriques dont l'irrégularité est supérieur à la dimension. *C. R. Acad. Sci. Paris* **212**, 1022–1024. [MR 6–102.]

Chevalley, C. (1951) *Introduction to the theory of algebraic functions of one variable.* Mathematical Surveys VI. American Mathematical Society, Providence, Rhode Island.

Coleman, R.F. (1985a) Effective Chabauty. *Duke Math. J.* **52**, 765–770.

Coleman, R.F. (1985b) p-adic abelian integrals and torsion points on curves. *Annals of Math.* **121**, 111–168.

Coleman, R.F. (1989) Torsion points on abelian étale coverings of $\mathbf{P}^1 - \{0, 1, \infty\}$. *Trans. Amer. Math. Soc.* **311**, 185–208.

Coombes, K.R. & Grant, D.R. (1989) On heterogeneous spaces. *J. London Math. Soc.* **40**, 385–397.

Coray, D. & Manoil, C. (1995) On large Picard groups and the Hasse Principle for curves and K3 surfaces. Preprint. Université de Genève.

Cremona, J.E. (1992) *Algorithms for modular elliptic curves*. Cambridge University Press.

de Jong, J. & Noot, R. (1991) Jacobians with complex multiplication. pp. 177–192 of *Arithmetic algebraic geometry (Texel, 1989)*. Progress in Math. **89**. Birkhäuser, Boston, Mass.

de Weger, B.M.M. (1992) A hyperelliptic equation related to imaginary quadratic number fields with class number 2. *J. reine angew. Math.* **427**, 137–156.

Donagi, R. (1980) Group law on the intersection of two quadrics. *Ann. Scuola Norm. Sup. di Pisa.* (4) **7**, 217–239.

Ein, L. & Shepherd-Barron, N. (1989) Some special Cremona transformations. *Amer. J. Math.* **111**, 783-800.

Faddeev, D.K. (1960) Group of divisor classes on the curve defined by $x^4 + y^4 = 1$. [In Russian.] *Dokl. Akad. Nauk SSSR* **134**, 776–777. [Translation: *Soviet Math. Doklady* **1**, 1149–1151.]

Fay, J.D. (1973) *Theta functions on Riemann surfaces*. Springer Lecture Notes in Mathematics, **352**. [MR 49 #569.]

Fischer, G. (1986) *Mathematical models, Commentary* (2 vols). Vieweg, Braunschweig.

Flynn, E.V. (1990a) The jacobian and formal group of a curve of genus 2 over an arbitrary ground field. *Math. Proc. Cambridge Philos. Soc.* **107**, 425–441.

Flynn, E.V. (1990b) Large rational torsion on abelian varieties. *J. Number Theory* **36**, 257–265.

Flynn, E.V. (1991) Sequences of rational torsions on abelian varieties. *Invent. Math.* **106**, 433–422.

Flynn, E.V. (1993) The group law on the jacobian of a curve of genus 2. *J. reine angew. Math.* **439**, 35–69.

Flynn, E.V. (1994) Descent via isogeny in dimension 2. *Acta Arith.* **66**, 23–43.

Flynn, E.V. (1995a) An explicit theory of heights. *Trans. Amer. Math. Soc.* **347**, 3003–3015.

Flynn, E.V. (1995b) A flexible method for applying Chabauty's Theorem. *Compositio Math.* To appear.

Flynn, E.V. (1995c) On a theorem of Coleman. *Manus. Math.* To appear.

Flynn, E.V., Poonen, B. & Schaefer, E.F. (1995) Cycles of quadratic polynomials and rational points on a genus 2 curve. Preprint.

Freije, M.N. (1993a) Kummer congruences and formal groups. *J. Number Theory* **43**, 31–42.

Freije, M.N. (1993b) The formal group of the jacobian of an algebraic curve. *Pacific. J. Math.* **157**, 241–255.

Frey, G. (1995) On elliptic curves with isomorphic torsion structures and corresponding curves of genus 2. pp. 79–98 of *Elliptic curves, modular forms, & Fermat's last theorem* (Eds. Coates, J. and Yau, S.T.). International Press, Boston, Mass.

Frey, G. & Kani, E. (1991) Curves of genus 2 covering elliptic curves and an arithmetical application. pp. 153–175 of *Arithmetic algebraic geometry* (Eds. van der Geer, G., Oort, F. and Steenbrink, J.). *Progress in Mathematics* **89**. Birkhäuser, Boston, Mass.

Geer, G. van der. *See* van der Geer, G.

Gonzales-Dorrego, M.R. (1994) (16, 6) configurations and geometry of Kummer surfaces in $\mathbb{P}^3$. *Memoirs Amer. Math. Soc.* # 512 (Vol. **107**), 1–101.

Gordon, D.M. & Grant, D. (1993) Computing the Mordell–Weil rank of jacobians of curves of genus 2. *Trans. Amer. Math. Soc.* **337**, 807–824.

Grant, D. (1988a) A generalization of Jacobi's derivatives formula to dimension 2. *J. reine angew. Math.* **392**, 125–136.

Grant, D. (1988b) Coates-Wiles towers in dimension 2. *Math. Ann.* **282**, 645–666.

Grant, D. (1990) Formal groups in genus 2. *J. reine angew. Math.* **411**, 96–121.

Grant, D. (1991) On a generalization of a formula of Eisenstein. *Proc. London Math. Soc.* **3**, 121–123.

Grant, D. (1993) Some product formulas for genus 2 theta functions. Preprint.

Grant, D. (1994a) Units from 3- and 4-torsion on jacobians of curves of genus 2. *Compositio Math.* **94**, 311–320.

Grant, D. (1994b) Integer points on curves of genus 2 and their jacobians. *Trans. Amer. Math. Soc.* **344**, 79–100.

Grant, D. (1994c) A curve for which Colman's effective Chabauty bound is sharp. *Proc. Amer. Math. Soc.* **122**, 317–319.

Grant, D. (1995) A proof of quintic reciprocity using the arithmetic of $y^2 = x^5 + 1/4$. *Acta Arith.* To appear.

Grant, D. *See also* Coombes, K.R. *and* Gordon, D.M.

Griffiths, P. & Harris, J. (1978) *Principles of algebraic geometry.* Wiley-Interscience, New York. [Especially Chapter 6.]

Harris, J. *See* Griffiths, P.

Hayashida, T. & Nishi, M. (1965) Existence of curves of genus 2 on a product of two elliptic curves. *J. Math. Soc. Japan* **17**, 1–16.

Heegner, K. (1952) Diophantische Analysis und Modulfunktionen. *Math. Zeit.* **56**, 227–263.

Heegner, K. (1956) Reduzierbare abelsche Integrale und transformierbare automorphe Funktionen. *Math. Ann.* **131**, 87–140.

Hodge, W.V.D. & Pedoe, D. (1952) *Methods of algebraic geometry* **II**. Cambridge University Press. [Especially pp. 232–234.]

Honda, T. (1968) Isogeny classes of abelian varieties over finite fields. *J. Math. Soc. Japan.* **20**, 83–95. [MR 37 #5216.]

Hudson, R.W.H.T. (1905) *Kummer's quartic surface.* Cambridge University Press. (Reprinted with commentary by W. Barth, CUP, 1990.)

Hulek, K., Kahn, C. & Weintraub, W.H. (1993) *Moduli spaces of abelian surfaces: compactification, degenerations, and theta functions.* de Gruyter expositions in mathematics, **12**. de Gruyter, Berlin.

Humbert, G. (1894) Sur les surfaces de Kummer elliptiques. *Amer. J. Math.* **16**, 221–253.

Humbert, G. (1899) Sur les fonctions abéliennes singulières. I, II, III. *J. Math. Pures Appl.* (Liouville) (5) **5** (1899), 233–350; **6** (1900), 279-386; **7** (1901), 97–123.

Humbert, G. (1901) Sur la transformation ordinaire des fonctions abéliennes. *J. Math. Pures Appl.* (Liouville) (5) **7**, 395–417.

Igusa, J.-I. (1960) Arithmetic variety of moduli for curves of genus 2. *Ann. of Math.* (2) **72**, 612–649. [MR 22 #5637.]

Igusa, J.-I. (1962) On Siegel modular forms of genus 2. *Amer. J. Math.* **84**, 175–200. [MR 25 #5040.]

Igusa, J.-I. (1964) On Siegel modular forms of genus 2, II. *Amer. J. Math.* **86**, 392–412. [MR 29 #6061.]

Igusa, J.-I. (1980) On Jacobi's derivative formula and its generalizations. *Amer. J. Math.* **102**, 409–446.

Inose, H. (1976) On certain Kummer surfaces which can be realized as nonsingular quartic surfaces in P3. *J. Fac. Sci. Univ. Tokyo, Sect. IA*, **23**, 544–560. [MR 55 #2924.]

Jacobi, C.G.J. (1832) Review of Legendre ' Théorie des fonctions elliptiques, troisième supplément.' *J. reine angew. Math.* **8**, 413–417. [= *Gesammelte Werke* **I**, 373–382.]

Jacobi, C.G.J. (1846) Über eine neue Methode zur Integration der hyperelliptischen Differentialgleichungen und über die rationale Form ihrer vollständigen algebraischen Integralgleichungen. *J. reine angew. Math.* **32**, 220–226. [= *Gesammelte Werke* **II**, 135–144.]

Jakob, B. (1994) Poncelet 5-gons and abelian surfaces. *Manus. Math.* **83**, 183–198.

Jessop, C.M. (1916) *Quartic surfaces with singular points*. Cambridge University Press. [Especially Chapter 9.]

Jong, J. de. *See* de Jong.

Kahn, C. *See* Hulek.

Kani, E. (1994) Elliptic curves on abelian surfaces. *Manus. Math.* **84**, 199–223.

Kani, E. *See also* Frey, G.

Keller, W. & Kulesz, L. (1995) Courbes algébriques de genre 2 et 3 possédant de nombreux points rationnels. *C. R. Acad. Sci. Paris, Série I*, **321**, 1469–1472.

Kempf, G.R. (1988) Multiplication over abelian varieties. *Amer. J. Math.* **110**, 765–773.

Klassen, M.J. & Schaefer, E.F. (1996) Arithmetic and geometry of the curve $y^3 + 1 = x^4$. *Acta Arith.* To appear.

Klein, F. (1870) Zur Theorie der Liniencomplexe des ersten und zweiten Grades. *Math. Ann.* **2**, 198–226.

Klein, F. (1926) *Entwicklung der Mathematik im 19. Jahrhundert*, Band I. Springer, Berlin.

Koblitz, N. (1989) Hyperelliptic cryptosystems. *J. Cryptology* **1**, 139–150.

Kouya, T. *See* Nagao, K.

Kramer, K. *See* Brumer, A.

Krazer, A. (1903) *Lehrbuch der Thetafunktionen.* Teubner, Leipzig.

Krazer, A. & Wirtinger, W. (1921) Abelsche Funktionen und allgemeine Thetafunktionen. *Enzyklopädie der Math. Wiss.* Band II, Teil 2, Heft 5, IIB7, 604–873. Teubner, Leipzig.

Kuhn, R.M. (1988) Curves of genus 2 with split jacobian. *Trans. Amer. Math. Soc.* **307**, 41–49.

Kulesz, L. (1995) Courbes algébriques de genre 2 possédant de nombreux points rationnels. *C. R. Acad. Sci. Paris, Série* I, **321**, 91–94.

Kulesz, L. *See also* Keller, W.

Kummer, E.E. (1864a) Über die Flächen vierten Grades mit sechszehn singulären Punkten. *Monatsber. Kön. Preuß. Akad. Wiss., Berlin* **1864**, 246–260. [= *Collected papers* **II**, 418–432.]

Kummer, E.E. (1864b) Über die Strahlensysteme, deren Brennflächen Flächen vierten Grades mit sechszehn singulären Punkten sind. *Monatsber. Kön. Preuß. Akad. Wiss., Berlin* **1864**, 495–499. [= *Collected papers* **II**, 433–439.]

Kummer, E.E. (1865) Über die algebraischen Strahlensysteme, in's Besondere über die der ersten und der zweiten Ordnung. *Monatsber. Kön. Preuß. Akad. Wiss., Berlin* **1865**, 288–293. [= *Collected papers* **II**, 441–447.]

Kummer, E.E. (1866) Über die algebraischen Strahlensysteme, in's Besondere über die der ersten und der zweiten Ordnung. *Abh. Kön. Akad. Wiss., Berlin* **1866**, 1–120. [= *Collected papers* **II**, 453–572.]

Lachaud, C. & Martin-Deschamps, M. (1990) Nombre de points des jacobiennes sur un corps fini. *Acta Arith.* **56**, 329–340.

Lang, S. (1959) *Abelian varieties.* Interscience, New York.

Lang, S. (1962) *Diophantine geometry.* Interscience, New York.

Lang, S. (1972) *Introduction to algebraic and abelian functions.* Addison-Wesley, Reading, Mass.

Lang, S. (1983) *Fundamentals of diophantine geometry.* Springer, New York.

Lange, H. (1987) Abelian varieties with several principal polarizations. *Duke Math. J.* **55**, 617–628.

Lange, H. (1991) Projective embeddings of abelian varieties. *Jber. Deutsch. Math.-Verein* **93**, 161-174.

Leprévost, F. (1991a) Famille de courbes de genre 2 munies d'une classe de diviseurs rationnels d'ordre 13. *C. R. Acad. Sci. Paris, Série* I, **313**, 451–454.

Leprévost, F. (1991b) Famille de courbes de genre 2 munies d'une classe de diviseurs rationnels d'ordre 15, 17, 19 ou 21. *C. R. Acad. Sci. Paris, Série* I, **313**, 771–774.

Leprévost, F. (1992) Torsion sur des familles de courbes de genre g. *Manus. Math.* **75**, 303–326.

Leprévost, F. (1993) Points rationnels de torsion de jacobiennes de certaines courbes de genre 2. *C. R. Acad. Sci. Paris, Série* I, **316**, 819–821.

Leprévost, F. (1995a) Jacobiennes de certaines courbes de genre 2: torsion et simplicité. Preprint.

Leprévost, F. (1995b) Sur une conjecture sur les points de torsion rationnels des jacobiennes de courbes. Preprint.

Leprévost, F. (1995c) Sur certains sous-groupes de torsion de jacobiennes de courbes hyperelliptiques de genre $g \geqslant 1$. Preprint.

Lesfari, L. (1988) Abelian surfaces and Kowalewski's top. *Ann. Sci. Ec. Norm. Sup.* **21**, 193–223.

Liu, Q. (1993a) Courbes stables de genre deux et leur schéma de modules. *Math. Ann.* **295**, 201–222.

Liu, Q. (1993b) Conducteur et discriminant minimal de courbes de genre 2. Preprint.

Liu, Q. (1994) Modèles minimaux des courbes de genre deux. *J. reine angew. Math.* **453**, 127–164.

Lutz, E. (1937) Sur l'équation $y^2 = x^3 - Ax - B$ dans les corps p-adiques. *J. reine angew. Math.* **172**, 237–247.

Manin, Yu.V. (1962) Two-dimensional formal abelian groups. [In Russian.] *Doklady Akad. Nauk SSSR* **143**, 35–37. [MR 26 #3772.]

Manoil, C. *See* Coray, D.

Manolade, N. (1988) Syzygies of abelian surfaces embedded in P4(C). *J. reine angew. Math.* **384**, 180–191.

Masser, D.W. & Wüstholz G. (1983) Fields of large transcendence degree generated by values of elliptic functions. *Invent. Math.* **72**, 407–464.

Mattuck, A. (1955) Abelian varieties over p-adic ground fields. *Ann. Math.* (2), **62**, 92–119. [MR 17-87.]

Mazur, B. (1977) Rational points of modular curves. Modular functions of one variable, V. pp. 107–148 of *Springer Lecture Notes in Mathematics*, **601**.

Mazur, B. & Swinnerton-Dyer, P. (1974) Arithmetic of Weil curves. *Invent. Math.* **25**, 1–61.

McCallum, W.G. (1988) On the Shafarevich-Tate group of the jacobian of a quotient of the Fermat curve. *Invent. Math.* **93**, 637–666.

McCallum, W.G. (1992) The arithmetic of Fermat curves. *Math. Ann.* **294**, 503–511.

Merriman, J.R. & Smart, N.P. (1993) Curves of genus 2 with good reduction away from 2 with a rational Weierstrass point. *Math. Proc. Cambridge Philos. Soc.* **114**, 203–214.

Mestre, J.-F. (1991a) Familles de courbes hyperelliptiques à multiplication réelle. pp. 193–208 of *Arithmetic algebraic geometry (Texel, 1989).* Progress in Math. **89**. Birkhäuser, Boston, Mass.

Mestre, J.-F. (1991b) Construction de courbes de genre 2 à partir de leurs modules. pp. 313–334 of *Effective methods in algebraic geometry* (Eds. Mora, T. and Traverso, C.). Progress in Math. **94**. Birkhäuser, Boston, Mass.

Mestre, J.-F. (1991c) Courbes elliptiques de rang $\geqslant 11$ sur $\mathbb{Q}(t)$. *C. R. Acad. Sci. Paris, Série I,* **313**, 139–142.

Mestre, J.-F. *See also* Bost, J.-B.

Milne, J.S. (1986a) *Arithmetic duality theorems.* Perspectives in Mathematics, **1**. Academic Press, Boston, Mass.

Milne, J.S. (1986b) Abelian varieties. pp. 105–150 of *Arithmetic Geometry* (Eds. Cornell, G. and Silverman, J.H.), Springer, New York.

Milne, J.S. (1986c) Jacobian varieties. pp. 167–211 of *Arithmetic Geometry* (Eds. Cornell, G. and Silverman, J.H.), Springer, New York.

Moret-Bailly, L. *See* Bost, J.-B.

Mumford, D. (1965) A remark on Mordell's conjecture. *Amer. J. Math.* **87**, 1007–1016.

Mumford, D. (1966) On the equations defining abelian varieties, I, II, III. *Invent. Math.* I, **1**(1966), 287–354; II, **3**(1967), 75–125; III, **3**(1967), 215–244. [MR 34 #4269, 36 #2621, 36 #2622.]

Mumford, D. (1974) *Abelian varieties.* Tata Institute, Bombay, and Oxford University Press.

Mumford, D. (1975) *Curves and their jacobians.* University of Michigan Press, Ann Arbor.

Mumford, D. (1983) *Tata lectures on theta,* I, II, III. Progress in Mathematics I, **28**(1983); II, **43**(1984); III, **97**(1991). Birkhäuser, Boston, Mass.

Murabashi, N. (1994) The moduli space of curves of genus two covering elliptic curves. *Manus. Math.* **84**, 125–133.

Nagao, K. & Kouya, T. (1994) An example of an elliptic curve over $\mathbb{Q}$ with rank $\geqslant 21$. *Proc. Japan Acad.* **70**, Series A, 104-105.

Nagell, T. (1935) Solution de quelques problèmes dans la théorie arithmétique des cubiques planes du premier genre. *Vid. Akad. Skrifter Oslo, Mat.-naturv. Kl.* **1935**, Nr 1, 1-25.

Namikawa, Y. & Ueno, K. (1973) The complete classification of fibres in pencils of curves of genus 2. *Manus. Math.* **9**, 143-186. [MR 51 #5595.]

Narasimhan, M.S. & Nori, M.V. (1981) Polarisations on an abelian variety. *Proc. Indian Acad. Sci. (Math. Sci.)* **90**, 125-128.

Narasimhan, M.S. & Ramanan, S. (1969) Moduli of vector bundles on a compact Riemann surface. *Annals of Math.* **89**, 14-51.

Nikulin, V.V. (1975) On Kummer surfaces. [In Russian.] *Izvestiya Akad. Nauk SSSR* **39**, 261-275. (Translation: *Math. USSR Izvestiya* **9** (1975), 278-293). [MR 55, #2926.]

Nishi, M. *See* Hayashida, T.

Noot, R. *See* de Jong, J.

Nori, M.V. *See* Narasimhan.

Ogawa, H. (1994) Curves of genus 2 with a rational torsion divisor of order 23. *Proc. Japan Acad.* **70**, Ser A, 295-298.

Ogg, A.P. (1966) On pencils of curves of genus 2. *Topology* **5**, 355-362. [MR 34 #1321.]

Oort, F. & Ueno, K. (1973) Principally polarized abelian varieties of dimension two or three are jacobian varieties. *J. Fac. Sci. Univ. Tokyo, Sect IA: Math.* **20**, 377-381. [MR 51 #520.]

Pedoe, D. *See* Hodge, W.V.D.

Poonen, B. *See* Flynn, E.V.

Ramanan, S. *See* Narasimhan, M.S.

Reid, M.A. (1972) *The complete intersection of two or more quadrics.* PhD Dissertation, Cambridge.

Richelot, F. (1836) Essai sur une méthode générale pour déterminer les valeurs des intégrales ultra-elliptiques, fondée sur des transformations remarquables de ces transcendantes. *C. R. Acad. Sci. Paris.* **2**, 622-627.

Richelot, F. (1837) De transformatione integralium Abelianorum primi ordinis commentatio. *J. reine angew. Math.* **16**, 221-341.

Rück, H.-G. (1990) Abelian surfaces and jacobian varieties over finite fields. *Compositio Math.* **76**, 351-366.

Ruppert, W.M. (1990) When is an abelian surface isomorphic or isogenous to a product of elliptic curves? *Math. Zeit.* **203**, 293–299.

Saito, T. (1989) The discriminants of curves of genus 2. *Compositio Math.* **69**, 229–240.

Schaefer, E.F. (1995) 2-descent on the jacobians of hyperelliptic curves. *J. Number Theory* **51**, 219–232.

Schaefer, E.F. (1996a) Class groups and Selmer groups. *J. Number Theory*. To appear.

Schaefer, E.F. (1996b) Descent on a jacobian using functions on a curve. In preparation.

Schaefer, E.F. *See also* Flynn, E.V. *and* Klassen, M.J.

Serre, J.-P. (1962) *Corps locaux*. Publications de l'Institut de Mathématique de l'Université de Nancago, VIII. Herrmann, Paris.

Serre, J.-P. (1988) *Lectures on the Mordell-Weil theorem*. Aspects of Mathematics, E15. Vieweg, Braunschweig.

Shepherd-Barron, *See* Ein, L.

Shimura, G. (1971) *Introduction to the arithmetic theory of automorphic functions*. Iwanami Shoten, Tokyo, and Princeton University Press.

Silverberg, A. (1988) Torsion points on abelian varieties of CM-type. *Compositio Math.* **68**, 241-249.

Silverman, J.H. (1985) Integral points on abelian varieties. *Invent. Math.* **81**, 341–346.

Silverman, J.H. (1986) *The arithmetic of elliptic curves*. Graduate Texts in Mathematics, **106**. Springer, New York.

Silverman, J.H. (1987a) Rational points on certain families of curves of genus at least two. *Proc. London Math. Soc.* **5**, 465–481. [MR 90a: 11067.]

Silverman, J.H. (1987b) Integral points on abelian surfaces are widely spaced. *Compositio Math.* **61**, 253–256. [MR 89c: 14070.]

Silverman, J.H. (1991) Rational points on symmetric products of a curve. *Amer. J. Math.* **113**, 471–508.

Smart, N.P. *See* Merriman, J.R.

Stoll, M. (1995) Two simple 2-dimensional abelian varieties defined over $\mathbb{Q}$ with Mordell-Weil rank at least 19. *C. R. Acad. Sci. Paris, Série* I, **321**, 1341–1344.

Swinnerton-Dyer, H.P.F. (1971) Applications of algebraic geometry to number theory. 1969 Institute on number theory. *Proceedings of Symposia in Pure Mathematics* **20**, 1–52. American Mathematical Society, Providence, Rhode Island.

Swinnerton-Dyer, H.P.F. (1974) *Analytic theory of abelian varieties.* LMS Lecture Notes, **14**. Cambridge University Press.

Swinnerton-Dyer, H.P.F. See also Mazur, B.

Terasoma, T. (1987) A Hecke correspondence on the moduli of genus 2 curves. *Comment. Math. Univ. St. Pauli* **36**, 87–115. [MR 88j: 14054.]

Top, J. (1988) A remark on the rank of jacobians of hyperelliptic curves over $\mathbb{Q}$ over certain elementary abelian 2-extensions. *Tôhoku Math. J.* **40**, 613–616.

Ueno, K. *See* Namikawa, Y. *and* Oort, F.

van der Geer, G. (1988) *Hilbert modular surfaces.* Ergebnisse der math. Wiss. (3), **16**. Springer, Berlin. [Especially Chapter 9.]

van der Geer, G. & van der Vlugt, M. (1992) Supersingular curves of genus 2 over finite fields of characteristic 2. *Math. Nachr.* **159**, 73–81.

van der Vlugt, M. *See* van der Geer.

Vlugt, *See* van der Vlugt.

Waterhouse, W. (1969) Abelian varieties over finite fields. *Ann. Sci. Ec. Norm. Sup.* **4**, 521–560. [MR 42 #279.]

Weger, B.M.M. de. *See* de Weger, B.M.M.

Weintraub, W.H. *See* Hulek, K.

Wüstholz, G. *See* Masser, D.W.

Zink, T. (1984) *Cartiertheorie kommutativer formaler Gruppen.* Teubner-Texte zur Mathematik **68**. BSB Teubner Verlagsgesellschaft, Leipzig. [MR 86j: 14046.]

Index rerum et personarum

basis elements (for jacobian) 6, 8
blowing down 6, 12
blowing up 6, 12
Brecht, B. 44
canonical class xii
canonical form 1
Castelnuevo's theorem 6, 12
complex multiplication 163
conjugate 2
defined (over k) xiii
degree (of a divisor) xi
desingularized Kummer 165 *et seq.*, 181, 184
diagonal 7
differential (on curve) xii
differential form 67
divisor xi
effective (divisor) xii
endomorphism ring 160 *et seq.*
even (basis element) 26
first kind (differentials of the) 4 *fn*
focus (of pencil of lines) 179
formal exponential map 68
formal group 10, 63 *et seq.*
formal logarithm 68
ftp 204
Galois module xiii
genus xii
Göpel tetrahedron 91
Grassmann coordinates 176
Grassmannian 175, 179
height 133 *et seq.*
Hilbert 90 xiii
homogeneous space 109
Hudson, R.W.H.T 17 *fn*
invariant differential 67
isogeny 88 *et seq.*, 101 *et seq*
jacobian 2, 6 *et seq.*

Jacobian 41 *et seq.*
kernel (of homomorphism μ) 53
kernel (of reduction) 69
Kummer (= Kummer surface) 17 *et seq.*
Littlewood's Principle 96
local parameters (for jacobian) 10, 28, 66
local uniformizer (on curve) xi
localized coordinates 63
Mordell-Weil group 47 *et seq.*
node 20
norm space 109
odd (basis element) 26
pencil (of lines) 175, 179
pencil (of quadratic forms) 177
Picard group xi
plane (of pencil of lines) 179
principal divisor xi
projective locus 8 *fn*
proper (automorph) 32 *fn*
quadric line complex 180
rational xiii
real multiplication 164
reducible (jacobian) 154 *et seq.*
reduction (map) 69
Richelot, F.J. 88 *fn*
Richelot isogeny 88 *et seq.*, 101 *et seq*
Riemann-Roch theorem xii
singular line (of quadric line complex) 180
support (of a divisor) xi
symmetroid 42
tetrahedroid 156
trope 25
twiddles 16
twist 109
wave surface 156
Weddle, T. 40 *fn*
Weddle's surface 40 *et seq.*
Weierstrass point xii

Printed in the United States
By Bookmasters